Security Implications of Climate Change for India

Published by Academic Foundation
in association with the
Institute for Defence Studies
and Analyses (IDSA), New Delhi

Security Implications of Climate Change for India

Report of the IDSA Working Group

ACADEMIC FOUNDATION
NEW DELHI

ACADEMIC FOUNDATION

First published in 2009
by

ACADEMIC FOUNDATION
4772-73 / 23 Bharat Ram Road, (23 Ansari Road),
Darya Ganj, New Delhi - 110 002 (India).
Phones : 23245001 / 02 / 03 / 04.
Fax : +91-11-23245005.
E-mail : books@academicfoundation.com
www. academicfoundation.com

in association with :

Institute for Defence Studies and Analyses, New Delhi
www.idsa.in

Cataloging in Publication Data--DK
Courtesy: D.K. Agencies (P) Ltd. <docinfo@dkagencies.com>

Security implications of climate change for India.

ISBN 13: 9788171887637
ISBN 10: 8171887635

1. Climatic changes--India--Congresses. 2. National security--Environmental aspects--India--Congresses. I. Institute for Defence Studies and Analyses (India).

DDC 551.60954 22

Typeset by Italics India, New Delhi.
Printed and bound in India.

Contents

About the Contributors 9
Foreword 13
Preface 19
Abbreviations 23

1. Climate Change and Security: Exploring the Link
   ARVIND GUPTA AND SUJIT DUTTA 27
2. Key Vulnerabilities Due to Climate Change
   MURARI LAL AND SUNIL CHAUHAN 43
3. Climate Change and Disasters
   SUNIL CHAUHAN 71
4. Climate Change and Migration
   P.K. GAUTAM 93
5. Energy and Climate Change
   SHEBONTI RAY DADWAL 105
6. Economics and Adaptation to Climate Change
   PURNAMITA DASGUPTA 117
7. Impact on India's Bilateral Relations with Neighbouring Countries
   UTTAM KUMAR SINHA, SREERADHA DATTA, SUNIL CHAUHAN AND P.K. GAUTAM 127

8. Impact on Warfighting Capability of the Indian Military
Sunil Chauhan, P.K. Gautam and Ajey Lele 143

9. Climate Change: India's Negotiating Position
Arvind Gupta, Sujit Dutta and Uttam Kumar Sinha 159

10. Conclusion
Arvind Gupta, Sunil Chauhan and P.K. Gautam 167

Appendices

Appendix A: Bali Action Plan 173

Appendix B: Indicative Timetable for Meetings of the Ad Hoc Working Group on Long-term Cooperative Action under the Convention in 2008 178

Appendix C: CO_2 Excluding LULUCF (Land Use, Land Use Change and Forestry) in Gg CO_2 Equivalent of some Annex 1 Countries in Giga Grams (1 Giga Gram=10 Billion Grams) 181

Glossary. 183

List of Tables and Figures

Tables

2.1 Climate Change Projections for Indian Subcontinent with Respect to Time under the SRES Marker Emission Scenarios . . . 51

2.2 Summary of Expected Effects on Extreme Events Over Indian Subcontinent . . . 64

4.1 Population Projections for India and Neighbouring Countries . . . 95

5.1 Top 20 Carbon Emitters . . . 107

5.2 Projections of CO_2 Emissions by Region . . . 108

5.3 Shares of Total GHG Emissions by Country/Region . . . 108

Figures

2.1 Observed Trends in All India Mean Surface Air Temperature during the 20th Century . . . 47

2.2 Observed Anomalies in All India Mean Summer Monsoon Rainfall during the 20th Century . . . 48

2.3 Observed Growth Rates in Global CO_2 Emissions as Compared to IPCC's SRES Marker Emission Scenarios . . . 49

2.4 The Contrasting Effects of Radiative Forcing due to Greenhouse Gases and the Aerosols on the Monthly Mean All India Mean Rainfall as Simulated by Climate Models . . . 52

2.5 Projected Changes in Monthly Mean Surface Air Temperature and Precipitation Averaged over India by the End of the 21st Century for A2 and B2 Scenarios . . . 54

3.1 Occurrence of Hydrometeorological Disasters: 1987 with Averages for Periods 1987-1998 and 2000-2006 . . . 73

4.1 Percentage Share of States in Total Projected Population Increase during 2001-2026 . . . 96

4.2 Population Projections . . . 98

About the Contributors

Sunil Chauhan was commissioned in the Indian Navy in 1985, and is currently a Research Fellow at the Centre of Strategic Studies and Simulation (CS3), USI, New Delhi. At CS3, his area of research is climate change, security and disaster risk management, with a focus on developing a new model/approach to disaster risk management, based on resilience, adaptability and transformability. He specialises in disaster and emergency risk management and besides having an MBA, is a Certified Master Trainer of Trainers from the Emergency Management Institute (EMI), Emmitsburg, USA. He has a number of certifications (specific to disaster/emergency management) including the Advanced Level Certification in Incident Command System (ICS) from the Lal Bahadur Shastri National Academy of Administration, Mussoorie, India and a Certificate in Humanitarian Assistance from UNHCR. He has published and presented a number of papers in national and international journals, seminars and workshops on climate change, disaster and emergency management issues and is a resource person for Indian institutes and other agencies for training on climate change, disaster management, conducting table top exercises and risk assessment.

Shebonti Ray Dadwal is currently a Research Fellow with the IDSA, specialising in energy-related issues. She has presented several papers in national and international seminars and conferences and has several peer-reviewed articles, projects and papers on issues related to energy security to her credit. In 2002, she published a book *Rethinking Energy Security in India* and is now writing her second book. Prior to joining IDSA, she worked as Senior Editor in *The Financial Express* and she has also served as Deputy Secretary at the National Security Council Secretariat.

Purnamita Dasgupta holds the Ford Foundation Chair and heads the Environmental Economics Unit at the Institute of Economic Growth, Delhi. A Ph.D. in environmental economics, she has been teaching and researching on environmental issues and their inter-relationship with economic development for almost two decades. In the recent past, positions held by her include: Member-Secretary, Net Present Value (NPV) Committee, Visiting Professor, University of Cambridge, and Member, Executive Committee, Indian Society for Ecological Economics (INSEE). She is well acquainted with the current debates on environmental sustainability, climate change and socio-economic issues in India. She has been researching on the interface between socio-economic scenarios and climate change in India for India's National Communication to the UNFCCC (NATCOM).

Sreeradha Datta is a Research Fellow at the IDSA since 1999. She is working on contemporary political and foreign policy related issues in South Asia. She has authored *The Northeast Complexities and its Determinants* and *Bangladesh: A Fragile Democracy* and co-edited *Changing Security Dynamics in Southeast Asia*. She also has a number of articles in reputed national and international journals besides edited chapters in a number of books. Earlier she was associated with the Centre for Peace Studies and edited *South Asia Watch* (1997-1999). She has given a number of talks at various think-tanks, security establishments and other academic institutions, and has been interviewed by various national and international media outlets including *BBC*, *Voice of America*, *The Economist*, *Time*, *Newsweek*, etc.

Sujit Dutta is a Senior Fellow at IDSA and works on China's foreign policy and diplomacy, state and politics in 20th century China, Asian security issues, India-China relations, international relations theory. He is currently engaged in a research project on "The Rise of China and its Impact on Asian Security". Prior to joining IDSA, he was a journalist with the Press Trust of India. He has been a member of the India-China Eminent Persons' Group set up by the two governments in 2001 for high-level Track-II dialogue on bilateral problems and diplomatic relations. He has represented IDSA and the India Committee at the Council for Security Cooperation in the Asia Pacific's Working Group on Confidence Building Measures. He is a frequent participant in national and international conferences. He has lectured and participated in seminars in Germany, China, Japan and the United States. He is also a regular speaker at India's leading international relations and strategic affairs institutions, including the National Defence College, the College of Warfare and the College of Naval Warfare. He leads the China and East Asia programme at the IDSA.

P.K. Gautam, a former Colonel of the Indian Army, is a Research Fellow since August 2005 at IDSA researching on environmental issues. He has authored: *Environmental Security* (2003), *National Security: A Primer* (2004), *Operation Bangladesh* (2007), *Composition and Regimental System of the Indian Army: Continuity and Change* (2008) and *Environmental Travelogue: Assessing and Projecting Future Role of the Military and Identifying New Security Challenges* (forthcoming).

Arvind Gupta joined the Indian Foreign Service (IFS) in 1979. He presently holds the Lal Bahadur Shastri Chair in Strategic and Defence Studies at the Institute for Defence Studies and Analyses. Prior to joining IDSA, Dr. Gupta was Joint Secretary at the National Security Council Secretariat (NSCS). During his tenure at the NSCS, he dealt with a wide spectrum of national security issues. He has wide-ranging diplomatic experience gained while working in Indian missions abroad. His current interests include: the impact of globalisation on India's security, the security, foreign policy and diplomatic challenges before India, energy security, climate change, institutional reform, India's neighbours, internal security and technology and security. He has several books, articles and papers to his credit.

Murari Lal currently, Chairman of Climate, Energy, Sustainable Development Analysis Centre (CESDAC) began his career as a Junior Scientific Officer at Indian Institute of Tropical Meteorology, Pune. He completed his Postdoctoral Fellowship in the Atmospheric Environment Service from Toronto. For over 20 years he had been teaching Environmental Physics and courses in Environmental Science at Indian and international universities/institutes. He specialises in global and regional/monsoon climate and its variability—scenario development, regional environmental change—integrated approach, sectoral vulnerability assessment (water, marine and coastal biodiversity, health and agriculture), landscape ecology, biophysical remote sensing—GIS applications, highland and coastal ecosystem modelling, EIA/SEA methods and practices, regional adaptation and mitigation approaches, carbon sequestration potentials and energy efficiency—clean development mechanism opportunities and options, natural resource management—technological options and interventions—environmental and economic feasibility, policy development including the review and analysis of multilateral environmental agreements in Asia—Pacific regional context, social dimensions and sustainable development. He has published a number of articles and reports in national and international scientific journals.

Ajey Lele is an Indian Air Force Officer and is currently working as a Research Fellow at the IDSA. He specialises on issues related to weapons of mass destruction (WMD) and space technologies. He is a postgraduate in Physics and Defence and Strategic Studies. He has obtained his doctorate form JNU, New Delhi. He has published many articles in journals, websites and newspapers. He has authored two books titled: *Bio-Weapons: The Genie in the Bottle* and *Weather and Warfare*.

Uttam Kumar Sinha is a Research Fellow at IDSA, working on non-traditional security issues. His areas of research include role of military in environment protection, human and water security. He is a member of the editorial board of *Strategic Analysis* and Assistant Editor of *Strategic Digest*. Besides, a doctorate from Jawaharlal Nehru University, he is a recipient of the Indo-Kuwait Dewan Award, 1995. Before joining the faculty at IDSA, he worked in the editorial of the national daily, *The Pioneer* and wrote the weekly column "Strategic Eye". He is a Visiting Faculty Member of the Malviya Centre for Peace Research, BHU and a Visiting Fellow to the International Peace Research Institute, Oslo. He was awarded the 2008 Indo-British Chevening 'Gurukul' Fellowship to the London School of Economics. Dr. Sinha is co-author of the book *Iraq War 2003: Rise of the New Multilateralism* and co-editor of the book *United Nations, Multilateralism and International Security*.

N.S. Sisodia took charge of IDSA in September 2005. He retired as Secretary in the Ministry of Finance in January 2005. Prior to this assignment, he was Secretary, Defence Production and Supplies, Ministry of Defence. He graduated from St. Stephen's College, Delhi and obtained a Master's Degree from Harvard University, USA, where he was a Mason Fellow. He joined the Indian Administrative Service in 1968 and served as District Magistrate in four districts of Rajasthan. He was later Principal Secretary, Industries & Commerce, Chairman and Managing Director of the State Industrial Development & Infrastructure Corporation and Chairman, State Electricity Board, Rajasthan. He has also been Vice Chancellor of Mohan Lal Sukhadia University, Udaipur. He is a member of the National Security Advisory Board.

Foreword

I would like to compliment the Institute for Defence Studies and Analyses for focusing attention on an extremely topical subject and I think this initiative is extremely timely. There is no doubt that we need much more research and analysis and more debate on the issues involved. We hope that this will be a continuing exercise and not just a one-off effort.

I would like to, first of all, make the point that at one level climate change is really an extremely pervasive phenomenon. It touches upon virtually every aspect of human existence. It is an overarching global issue. It impacts both our national security as well as global peace and security in general. There is also another level we should look at. Even though climate change has a global impact, this will be asymmetrical in terms of regions and in terms of countries which are impacted. In some countries the impact may be more severe, in others less. This may involve taking into account geographical location, for example, whether one is in the temperate zone or the tropical or the equatorial zone. It may be related to physical terrain features, for example, whether the state is a coastal or an island state, a peninsular state or whether it is a country with high altitude mountains. The economic capacity and resilience of the countries or regions which are impacted and their resilience is related to income levels. They differ for a poor country compared to a rich country. There are also factors relating to the institutional structures. Well-developed institutions can take care of the impact of climate change. An institutional structure for proper surveillance and monitoring of the impact of climate change can help adaptation. All these factors have to be taken into account when security implications of climate change are studied. This means that even though we say climate change is a global phenomenon and has a global impact, we have to look at regional and national impacts because these are conditioned by the above variables and peculiar national circumstances.

In this respect, India and the region it is situated in, has its own particularity and specific challenges. First, there is ongoing environmental degradation which is already taking place. What this means is that even if we do not take into account climate change related stresses, we have environmental problems. Virtually all the stresses that have emerged in the past several years due to high population growth, uneven economic growth and physical and climatic diversity of this region, will get much more accentuated as a result of climate change.

By contrast, some of the countries which are situated in the temperate zone feel that in the short-term they may even gain due to climate change because of milder temperatures, milder winters and longer crop-growing seasons. The melting of the Arctic is already having a geopolitical and economic impact. The melting of the Arctic has now allowed some commercial movement of ships from Japan to Northern Europe using the northern route for the very first time. It is likely that within the next few years this will become a major sea route. This will mean that the whole centre of gravity of global shipping lanes will shift, which will have a major impact on the global economy. Added to that of course, is the fact that about 25-30 per cent of the current fossil fuel reserves in the world may be in the Arctic. Some initial exploration is already taking place. There is a bit of a scramble amongst the countries who claim various chunks of the Arctic. These countries have decided to come together as a group to try and moderate their conflicts and make sure that this does not spillover into any kind of armed confrontation. It would be very worthwhile doing a case study of the melting of the Arctic itself and the responses of the countries around that area because these may have very important lessons for us. What is clear is that there will be asymmetrical impacts in terms of climate change and the ability of various countries to adapt to this climate change.

For India, we have to proceed on the basis of some key assumptions. These are important not only in terms of what we do nationally but also in terms of our international negotiating position. First of all, even if greenhouse gas emissions were halted entirely, climate change will continue to take place. This is because climate change is not taking place because of the current emissions of various countries but due to the impact of accumulated emissions in the planetary atmosphere—accumulation which has been taking place through the process of carbon-based industrialisation for the last 200 years. We should not be deflected by the very narrow focus on current emissions, which is what the developed countries are trying to do. This is not based on science. This is extremely important because there is such a high volume and high decibel campaign to focus attention on current emissions that this particular dimension of climate change has actually been lost. We should be clear about the facts; and the facts are that even if tomorrow there is a complete halt to greenhouse gas emissions, climate change will continue to take place and the burden of adjustment to this climate change will continue to be shouldered by countries like India.

India has to engage in sustainable development. All our national policies need to factor in climate change. It is in our interest that we move from a strategy of growth which is based on carbon and fossil fuels to a strategy of growth which is based on renewable sources. This is because even if theoretically there was no climate change, in order to maintain 8-9 per cent growth, our energy constraint will have to be dealt with. Once we get over the current global economic and financial crisis, energy will again emerge at the top of the international agenda. We are living in a world of depleting fossil fuels and there is no way that we can sustain the kind of growth rates that we are looking for if we continue to depend on such fuels. So, in any case we have to make a shift to renewables. Nevertheless, whatever we do nationally is going to be conditioned and impacted by what everybody does together. India is mindful of the fact that national action is taking place not in isolation but within the context of a global phenomenon. We on our own, may do very drastic things in response to climate change but this may mean nothing at all if the global environment continues to deteriorate. There is an integral link between what we do nationally and the negotiation of a global compact. This global regime must be supportive of, and be conducive to, the effective implementation of the measures which are there in our National Action Plan. This is why the current multilateral negotiations are so important because it is in our vital interest that we work together. There may be shifting coalitions. It is not possible to predict that all the developing countries, irrespective of their situations, will band together on every issue. We may be able to develop different kinds of coalitions. The aim of building these coalitions should be to ensure that the global compact which emerges is something which is fair and equitable. It needs to be supportive of the kind of strategy that we ourselves are adopting for sustainable development. This global dimension has to be factored in.

Another important issue relates to our own plans for adaptation. 'Development is the best form of adaptation' is not just a slogan. Development builds climate resilience. Adaptation, in the final analysis, is building a more resilient economy, a more resilient institutional structure which enables us to deal with the kind of impacts that we expect. I am happy to note that these possible impacts have been very well spelt out in the Report.

Finally, the link between climate change and national defence is very critical. The impact of climate change is pervasive on virtually every aspect of our national life including national defence. This requires a very specific and specialised study. For example, it has been mentioned in the study how the melting of Himalayan glaciers will impact on our force deployments in these mountainous areas. Since a major part of our deployments are in high altitude areas, climate change will lead to changes in terrain, requiring reorientation of our military tactics. Again, a reference was made to our naval assets and how they would be impacted. We have the example of the Tsunami. This was an example of the kind of extreme climatic or natural disaster which may result from climate change. Vulnerabilities of coastal areas, vulnerabilities of island territories, all these are things which will impact on the manner in which the Indian Navy determines what its doctrine should be, what its objectives should be and what is the best kind of structure of our naval forces required to deal with these challenges. A reference in the study has been made to the Arctic Ocean and I think we need to look at what this means in terms of global naval strategy.

A reference has also been made to what will happen in terms of our relationship with our neighbours. All the environmental stresses which I mentioned are also stresses which are operating on our neighbouring countries. Therefore, in a sense, we are in it together and we have to try and see how we can develop more cooperative structures to deal with common threats. We should not allow an escalation of conflict situations as a result of climate change. Of course, there is also the aspect of internal security. Internal security may also be impacted by the increasing frequency of extreme climatic events, the frequency of natural disasters and large movements of population. The conclusion is obvious. We need capacity building to deal with these kinds of events. We need a very good system of coordination amongst different agencies and different ministries to cope with such new challenges. I would take as an example the recent Kosi disaster and the impact of the floods on a very large population living in the Indo-Gangetic plains. This points to the need for good monitoring systems. We also need a system to quickly respond to such events which will, in all likelihood, increase in frequency and intensity as climate change progresses.

The IDSA study has initiated a good process of dialogue and debate on critical issues of national security for India. We should evolve an Indian perspective on climate change and national security.

— Shyam Saran

Special Envoy to the Prime Minister

New Delhi

Preface

Climate change is no longer just a matter of scientific or intellectual curiosity. Over the years, it has come to be regarded as one of the most potent existential threats to civilisation. There may be disagreements about the precise extent, scope, intensity and consequences of climate change but there is general consensus that climate change poses a danger to humanity and it needs to be tackled urgently.

It is in the backdrop of this grave challenge posed by climate change that the IDSA undertook the present study to explore the links between climate change and security. IDSA's core competence is in security issues and not in climate change *per se*. Yet, a realisation has been growing that through various direct and indirect inter-relationships among physical, economic and social phenomena, climate change will have a discernible impact on national and international security. To explore the issue, IDSA organised a national workshop in April 2008 on the subject of "Climate Change and National Security". A number of renowned experts made their presentations. It was felt that the link between climate change and national security needed to be probed in greater depth. It was, therefore, decided to set up a Working Group comprising the following experts and researchers to examine the issue further:

From IDSA

• Dr. Arvind Gupta, Lal Bahadur Shastri Chair	Chairperson
• Shri Sujit Dutta, Senior Fellow	Member
• Col. P.K. Gautam (Retd.), Research Fellow	Member
• Dr. Uttam Sinha	Member
• Wg. Cdr. Ajey Lele, Research Fellow	Member
• Ms. Shebonti Ray Dadwal, Research Fellow	Member
• Dr. Sreeradha Datta, Research Fellow	Member
• Shri T. Khurshchev Singh, Research Assistant	Member
• Ms. Monalisa Joshi, Research Assistant	Member

EXTERNAL EXPERTS

- Cdr. Sunil Chauhan (IN), Research Fellow, USI
- Prof. Murari Lal, Chairman, Climate, Energy, Sustainable Development Analysis Centre
- Prof. Purnamita Dasgupta, Head, Ford Foundation Chair and Head, Environmental Economic, Unit, IEG

The Working Group members worked hard and with extraordinary dedication for preparation of the Report. They provided the initial drafts of various chapters and also contributed to the development of the Report as a holistic document.

A number of working papers were prepared by the Working Group members. These were discussed at length at the meetings. The final Report is the outcome of the Group's deliberations. This Report on the Working Group's findings is presented for dissemination and public debate.

The Report has attempted to take into account the latest available information on climate change and assess its impact on India as well as the subcontinent. The findings are being presented in a non-technical and simple language so that they are accessible to the wider public.

The Report does not get into details on every aspect of climate change. That was not its purpose. The Working Group has attempted to focus on the key vulnerabilities arising out of climate change, which have a bearing on India's security as broadly construed.

The Report assigns a wider meaning to "national security". The elements of human security are considered to be an integral part of the concept of "national security". Climate change is not about wars. It is more about gradual build-up of stresses, tensions and sometimes violent conflicts due to the breakdown of physical and environmental systems that are crucial for life.

There has been considerable debate on whether or not climate change should be 'securitised' i.e., whether or not climate change should be treated as a hardcore security problem to be tackled by international security institutions like the UN Security Council under Chapter VI and VII of the

UN Charter and through military instruments. The obvious answer, from the developing country perspective, is "No". The Report agrees with the view that to be unaware of the security dimensions of climate change will be counter-productive. There is a difference between the security dimension of climate change which undoubtedly manifests through various pathways and 'securitisation' of climate change, which is a pressure tactic being used by some countries to serve their own agenda.

The Report has reached a few conclusions which are presented in Chapter 10. The idea is not to delve deep into the basics of negotiating positions of the Government but to identify the areas where our stances may have to evolve in view of future pressures, both domestic and international.

The Working Group Report does not aim to offer detailed normative advice to the Government. That was not the purpose. It, however, states categorically that there are a variety of challenges to national security, arising from climate change, which should be acknowledged. Acknowledging these challenges does not mean that the Government should support 'securitisation' of climate change. Yet, it should be ready with its responses if the others succeed in securitising the issue.

One may ask, what is new or fresh in the Report? The Report seeks to bring together at one place the latest available information on climate change in the Indian context. It identifies India's key vulnerabilities, examines the climate change issues in international and domestic context, alerts the reader about the link between climate change and national security and makes a few recommendations as to how India's foreign and socio-economic policies should factor in climate change right from the beginning. It makes the point that India, without taking any binding commitment on emission reduction, should nevertheless, consider formulating target-oriented policies to improve energy efficiency, popularise clean technologies, etc. Climate change should be used as an opportunity to make socio-economic development in India more efficient and sustainable.

The Report was discussed at a Round Table of experts held at IDSA on 27 November 2008 under the chairmanship of Shri Prodipto Ghosh. Shri Shyam Saran, Prime Minister's Special Envoy, delivered the keynote address. The Report was critiqued by Shri Nitin Desai, Shri C. Dasgupta, Shri Prodipto Ghosh, Shri Manjiv Puri and Maj. General (Retd.) K.B. Kapoor, VSM and several other participants. The comments of the discussants have been taken on board wherever necessary. The IDSA thanks Shri Shyam Saran for writing the Foreword to the Report. Our thanks are also due to the discussants for their insightful comments.

I would like to thank the members of the Working Group for their commendable effort in developing this Report. I also acknowledge the valuable research assistance provided by the young researchers at the IDSA.

— N.S. Sisodia

Director General, IDSA

New Delhi

Abbreviations

AOGCM	Atmosphere Ocean Global Climate Models
AR	Assessment Report
BDCLIM	Bangladesh Climate
BEE	Bureau Energy Efficiency
BIMARU	Bihar, Madhya Pradesh, Rajasthan and Uttar Pradesh
BMCTP	Building Material Council and Technology Promotion
BNP	Bangladesh Nationalist Party
CCDRI	Climate Change/Disaster Risk Indices
CCS	Carbon Capture and Storage
CCSR	Center for Complex Systems Research, Japan
CCT	Clean Coal Technology
CDM	Clean Development Mechanism
CERs	Certified Emission Reductions
CESDAC	Climate, Energy, Sustainable Development Analysis Centre
CNA	Center for Naval Analyses
CO_2	Carbon Dioxide
CRED	Centre for Research on the Epidemiology of Diseases
CS3	Centre of Strategic Studies and Simulation
CSIR	Council of Scientific and Industrial Research
DCDC	Development, Concepts and Doctrine Centre
DEFRA	Department for Environment, Food and Rural Affairs
DRI	Disaster Risk Index

ECA	Energy Conservation Act
EEZ	Exclusive Economic Zone
EMI	Emergency Management Institute
ETS	Emissions Trading Scheme
EU	European Union
FICDM	Fellow of the International Congress on Disaster Management
FIETE	Institution of Electronics and Telecommunication Engineers
FSU	Former Soviet Union
G8	Group of Eight nations—Canada, France, Germany, Italy, Japan, Russia, The United Kingdom and The United States
GCM	General Circulation Models
GDIN	Global Disaster Information Network
GDP	Gross Domestic Product
GHG	Greenhouse Gas
GLOF	Glacial Lake Outburst Floods
GoI	Government of India
GtC	Gigatonnes of Carbon
HGS	High Growth Scenario
IAEM	International Association of Emergency Managers
IARI	Indian Agricultural Research Institute
ICS	Incident Command System
IDSA	Institute for Defence Studies and Analyses
IEA	International Energy Agency
IGCC	Integrated Gasification Combined Cycle
IIM	Indian Institute of Management

IISc	Indian Institute of Science
IIT	Indian Institute of Technology
IMD	India Meteorological Department
INSEE	Indian Society for Ecological Economics
IPCC	Inter-governmental Panel on Climate Change
IVCF	Indian Venture Capital Firm
IWT	Indus Water Treaty
JRC	Joint Rivers Commission
KWh	Kilowatt hour
LULUCF	Land Use, Land-Use Change and Forestry
MDG	Millennium Development Goals
MEA	Ministry of External Affairs
MoD	Ministry of Defence
MoEF	Ministry of Environment and Forests
MoST	Ministry of Science and Technology
NAPCC	National Action Plan on Climate Change
NATCOM	India's Initial National Communication
NDMA	National Disaster Management Authority
NGO	Non-Governmental Organisation
NIES	National Institute for Environmental Studies, Japan
NIO	National Institute of Oceanography
NPL	National Physical Laboratory
NPV	Net Present Value
OECD	Organisation for Economic Co-operation and Development
PACE-SD	Pacific Centre for Environment and Sustainable Development
PGM	Precision Guided Munitions

PMO	Prime Minister's Office
PRECIS	Providing Regional Climate for Impact Studies
R&D	Research and Development
RCM	Regional Climate Models
RUSI	Royal United Services Institution
SAARC	South Asian Association for Regional Cooperation
SLORC	State Law and Order Restoration Council
SLR	Sea Level Rise
SNC	Second Indian Nation Communication
SRES	Special Report on Emissions Scenarios
TERI	The Energy and Resources Institute
UN	United Nations
UNCLCS	United Nations Commission on Limitation of Continental Shelf
UNCLOS	United Nations Convention on the Laws of the Sea
UNDP	United Nations Development Programme
UNEP	United Nations Environmental Programme
UNFCC	United Nations Framework Convention on Climate Change
UNFC	United Nations Framework Convention
UNSC	United Nations Security Council
US	United States
WBGU	German Advisory Council on Global Change
WEO	World Energy Outlook
WMO	World Meteorological Organization

1 Climate Change and Security

Exploring the Link

Arvind Gupta and Sujit Dutta

Is there a link between climate change and national security? This chapter seeks to explore this important question.

Climate Change is Now an Accepted Fact

Until a few years ago there was a lot of scepticism about the reality of climate change. The scientists, who have been studying the phenomenon of global warming for several years, began to establish the link between climate change and anthropogenic emissions of greenhouse gases in the seventies and eighties. The first Climate Conference (Geneva, 1979) organised by the UN Environmental Programme (UNEP) and World Meteorological Organization (WMO) was formed on climate change. Several scientific workshops were held in the eighties, including those in Villach (Austria, 1987) and Bellagio (Italy, 1987) under UNEP/WMO auspices. Recognising the importance of the issue, WMO established an International Panel on Climate Change in 1988 to study the problem further and present a scientific consensus on the subject. The IPCC produced its first report in 1990, the second in 1995, the third in 2001 and the fourth in 2007. These reports presented the scientific consensus on climate change and were meant for a wide community of policymakers, scientists, experts in various fields as well as for wider publicity. The first report established a link between global warming and the emission of greenhouse gases. It also drew attention to the consequences of climate change and stressed the need for urgent action on the part of governments. The report was greeted with scepticism, particularly by politicians and governments. Their main argument was that there was considerable uncertainty about the phenomenon of climate change itself. It was pointed out that there was no scientific consensus on several critical aspects of global warming. Critics feel that the projections about the impacts and consequences were mere speculations. The governments were reluctant to commit themselves to costly pre-emptive action on a phenomenon which appeared to be uncertain.

The situation began to change gradually with the first Earth Summit in Rio in 1992. The governments were able to agree on a convention on climate change, the UN Framework Convention on Climate Change (UNFCC) in 1992. In a landmark development, the Kyoto Protocol on emissions cut was signed in 1997. It came into force in February 2005. Carbon trading began in 2005. Discussions have begun on commitments after 2012. The agreement is expected to be reached in Copenhagen in December 2009.

The scepticism about climate change has reduced considerably even though the uncertainty about the phenomenon and its consequences has not been removed entirely. An increasing number of people now believe that climate change is a reality and something should be done about it while there is still time on the side of humanity. The anxiety about climate change has been further aggravated by the palpable rise in recent years in the frequency and spread of extreme weather phenomena like heat waves, hurricanes, floods and droughts, melting of glaciers, etc. The media and environmental NGOs have also heightened the public fear about the adverse consequence of climate change by sensationalising the issue. It may be added here that while scientific consensus on the phenomenon of climate change has been improving steadily over the last two decades, there are vast differences even among experts, policymakers and academics on what is the best way forward to deal with the problem. The international negotiations on climate change are marked by sharp differences between the developed and industrialised countries and the developing countries. The Stern Review (2006), which dwelt upon the economics of climate change, has been widely discussed but its findings based on integrated assessment models, have been criticised by experts, including those in India.

During the Cold War years nuclear weapons were regarded as the foremost threat to humanity. Climate change is considered by many as the "existential" threat—number one. A huge amount of international attention is focused on climate change, its adverse consequences and impacts and how to mitigate them.

Impact of Climate Change: IPCC's Projections

Our present state of knowledge, imperfect though it is, is embodied in the Fourth Assessment Report (AR4) of the IPCC published in December 2007. The authoritative 19-page Summary for Policymakers issued by the IPCC presents the prevailing consensus among experts on the observed changes in climate and their effects, causes of change, projected climate change and its impact, adaptation and mitigation options, and the long term perspective.[1] AR4 considers six GHG emission scenarios for global averaged surface warming and sea level rise up to the end of the 21st century. These emission scenarios range from the mildest (B1) to the most severe (A1 F1). For the B1 scenario, the best estimate for mean temperature rise is 1.8°C as compared to 4°C for A1 F1 scenario. The mean sea level rise projected for the B1 scenario is in the range of 0.18-0.38 metres and in the range of 0.26-0.59 metres for A1 F1 scenario.[2]

While the surface mean temperature of the planet is expected to rise under all scenarios, there will be considerable regional variations. The warming is expected to be the greatest in the high Northern latitudes and the least over the Southern ocean and parts of the Northern Atlantic Ocean. There will be a contraction of the snow-covered area with the near complete disappearance of late summer sea ice in the Arctic. This is already happening. The frequency of heat waves and heavy precipitation, according to AR4, will "very likely increase". There is a likelihood of an increase in tropical cyclonic intensity. Precipitation would increase in high latitudes and decrease in sub-tropical land regions. Annual river run-off and water availability will increase at high latitudes and in some tropical wet areas, while the same will decrease in the mid-latitudes and tropics. The Mediterranean basin, Western United States, Southern Africa and North East Brazil will experience water shortages due to climate change.[3]

The report gives several examples of the impacts associated with global warming. These pertain to water, ecosystems, food, coasts and health. It provides examples of regional impacts on Africa, Asia, Australia and New Zealand, Europe, Latin America, North America and Polar

1. Intergovernmental Panel on Climate Change, Fourth Assessment Report, Climate Change 2007: Synthesis Report, Summary for Policymakers.
2. Ibid., Table, SPM 1.
3. Ibid.: 8.

regions.[4] AR4 also provides examples of the impact of climate change on various sectors, namely, agriculture, forestry and ecosystems, water resources, human health and industry and settlement and society.[5]

Some of the projections are serious. For example:[6]

- Africa: By 2020, about 75-250 million people in Africa will be exposed to water stress due to climate change; in some countries agricultural production will be reduced by up to 50 per cent; fisheries resources will deplete in large lakes; the low-lying coastal areas with large populations will be affected by the reduction in fisheries resources. Africa will be one of the most vulnerable continents due to climate change.
- Asia: By 2050, fresh water availability in Central, South, East and Southeast Asia will decrease. Coastal areas will be exposed to greater risk of increased flooding; the melting of glaciers in the Himalayas will lead to increased flooding within 2-3 decades. Thereafter, river flow will decrease. Climate change will adversely affect sustainable development in most countries in Asia. Crop productivity will decrease by about 20 per cent in East and Southeast Asia and by 30 per cent in Central and Southeast Asia by the mid-21st century. There will be a rise in diarrhoeal diseases due to floods and droughts as well as changes in the hydrological cycles connected with global warming.
- Australia and New Zealand: Reduced precipitation, significant loss of biodiversity, increase in the severity and frequency of storms and coastal flooding by 2050 is projected. Agriculture and forestry yields are also likely to decline by 2030.
- Europe: Europe will experience wide-ranging effects of climate change including retreating glaciers, longer growing seasons and heat waves. Regional differences in Europe will increase. In southern Europe, water availability will reduce, affecting hydropower potential, summer tourism and crop productivity.

---

4. Ibid., Table SPM 2, p.10.
5. Ibid., SPM 3, p.12.
6. Based on Table SPM 2 of AR4 Summary for Policymakers, p.10, and Working Group 2 contribution to the Fourth Assessment Report.

- Latin America: Tropical forests will reduce in Eastern Amazonia. Variations and precipitations and the disappearance of glaciers will affect water availability. Rise in sea level will cause increased flooding in low-lying areas.
- North America: Decrease in snow-packs, reduced summer flows, more winter flooding, heat waves, pressure on coastal communities and habitats are projected for North America.
- Polar regions: Reduction in thickness of ice sheets and sea ice will be the major impact of climate change in the Polar regions.
- Small islands: Erosion of beaches, coral bleaching, decline in fisheries stocks, inundation and strong surge due to rise in sea level are expected to pose major risks to small islands. Water availability in the Caribbean and Pacific islands will reduce.

Socio-Economic Impact of Climate Change

What will be the socio-economic impact of such changes? Large-scale disruptive migrations of affected populations in search of food, water and safer locations will take place. Such migrations will change demographic patterns, create further scarcities in the areas to which the migrants move, strain the existing resource base and trigger fresh tensions. Rise in temperatures will also lead to the outbreak of pandemics which would spread across the globe.

While the entire planet will be susceptible to the adverse impact of climate change, the poorer regions will be affected more severely than the affluent ones. This is because the poorer developing countries' capacity to adapt to adverse impacts is low. They do not have the financial and technical resources to deal with the problems on such a large scale as envisaged under the various IPCC scenarios.

Many developing countries are already battling with socio-economic crises of various kinds. Climate change will accentuate these crises. Some of them, like the Maldives Kiribati or Tuvalu face serious risk of submergence due to the rise in sea levels. For them, climate change is an existential threat. It is interesting to note that Kiribati and Tuvalu have already requested Australia and New Zealand to accept the relocation of the Island's entire population in case of emergency.

Climate change may aggravate the existing conflicts and further increase the rich-poor divide in the world. This will have adverse implications for the emergence of an equitable international order. One of the most dramatic illustrations of the impact of climate change is provided by the rapid melting of the Polar ice sheets and mountain glaciers. Almost 99 per cent of the world's ice is located in the Arctic and Antarctic ice sheets and the rest in the mountain glaciers. The ice sheets of the Arctic and Antarctic are already melting. They may begin to melt at an accelerated pace as the global mean temperature rises.[7] If they melt completely, the sea level could rise by over 70 metres. Several glaciers are in retreat. The glacier atop Mount Kilimanjaro in Africa has already lost 90 per cent of its mass. The Himalayas possess one of the largest resources of snow and ice and its glaciers form a source of water for perennial rivers such as the Indus, the Ganga and the Brahmaputra. Melting of the glaciers may impact their long-term lean season flows with adverse effects on the economy in terms of water availability and hydropower generation.[8] The disappearance of the Himalayan glaciers, predicted in some scenarios to take place by 2050, will result initially in floods and subsequently water shortages for the millions of people of South Asia. Water shortages would trigger food and water scarcities and a host of other problems.

The rise in global mean temperature would result in increase in vector-borne diseases like malaria and dengue. These two diseases have already spread to areas which were earlier free of them. Higher incidence of natural disasters like floods and cyclones will lead to higher casualties, thereby putting greater strain on public health systems. Poorer countries that have inadequate public health systems will have to bear the brunt of the increased risk to human and animal health due to climate-related factors.

Agriculture is highly sensitive to rainfall patterns and temperature. A 1-3 degree celsius rise in the next 50 years will reduce rice, maize and wheat yields in most parts of the world, particularly, Africa, South Asia, Southeast Asia and Central Asia in varying degrees. Reduction in crop yields will create food shortages and socio-economic problems including conflict situations. The increasing incidence of drought will aggravate the

7. Bell, Robin E. (2008). "The Unquiet Ice", *Scientific American India*, February, pp.50-57.

8. Government of India (2006). *National Action Plan on Climate Change*. Prime Minister's Council on Climate Change. June. p.15.

food shortages even more. Higher temperatures will also affect the world's coral reefs, mangroves and the associated ecosystems. This in turn will reduce fish stocks greatly.

Impact on Geopolitics

Even if we do not agree with the worst case scenarios or dire projections of climate change for the planet, it is reasonable to expect that climate change will have unpredictable geopolitical consequences. One striking example is provided by the melting of ice caps in the Arctic Ocean. The ice in the Arctic Ocean surrounding the North Pole has thinned down drastically in the last few decades. Global warming is considered to be the cause. The melting of the Arctic ice has major geopolitical implications. For the first time in human history, the Northwest Passage connecting Europe with Asia and North America will open up resulting in new shipping and trading routes. The melting of the Arctic ice will also make vast areas in the continental shelf of the five Arctic Circle countries (the US, Canada, Denmark, Norway and Russia) accessible for exploitation of oil, natural gas and numerous other minerals. So important is the opening of the Arctic Ocean that Russia has staked its claim with the UN Commission on Limitation of Continental Shelf (UNCLCS) on 1.19 square kilometres of the Continental Shelf extending from Siberia. In August 2007, a Russian expedition, in a highly symbolic gesture, planted a Russian flag on the sea floor 4 kilometres below the North Pole. The Russian Strategic Aviation bombers have carried out numerous flights over the North Pole. This has evoked sharp criticism from the other Arctic Circle countries. New disputes amongst the Arctic Circle countries have arisen over territory and resources. These countries are making their own plans to enhance their presence and claims in the region. Canada, Denmark, the US and Russia are refurbishing their navies and air forces for future operations in the Arctic region. These events have led some analysts to warn of the possibility of "armed conflicts" in the region.[9] The Arctic countries are prone to think of the Arctic region as their backyard where other countries have no rights. While the Antarctica is governed by the Antarctica Treaty which is an international agreement, the Arctic region countries oppose

9. Borgerson, Scott G. (2008). "Arctic Meltdown: The Economic & Security Implications of Global Warming", *Foreign Affairs* 87(2): 66-77, March/April.

any such arrangement for the Arctic region. The Arctic provides a dramatic illustration of the unpredictable geopolitical cover of climate change. Another example of geopolitical change due to global warming is that of the melting of permafrost on the Tibetan plateau. While there would be some negative impacts of climate change like the melting of the Himalayan glaciers, the thawing of the permafrost could make available more arable land to China and trigger the migration of the Han population to Tibet in even greater numbers. A reordering of the international order on the basis of decarbonisation of global economy is likely to be expected.

The Key Issue: Dependence on Fossil Fuel

Climate change is intimately related to energy use patterns. The CO_2 emissions caused by the burning of fossil fuels are the main culprit behind climate change. Thus, any effort to reduce emissions of CO_2 means reducing the burning of fossil fuel. In the absence of viable non-carbon energy sources, the reliance on fossil fuel for energy is unlikely to reduce at least in the near to medium term. The search for more non-carbon sources of energy like solar and wind energy has begun but the results are as yet not promising. Renewable energy is as yet uneconomical. Nuclear energy is seeing a revival but the concerns about non-proliferation and the anxiety about nuclear material falling into the hands of terrorists have also increased.

The former US Vice-President Al Gore has called for the generation of the entire electricity in the US from non-fossil fuel sources in the next 10 years. This is a radical proposal meant to push climate change higher in policy priorities of the US. The Obama Administration is expected to give a higher priority to climate change than the Bush Administration ever did. This is likely to have a major impact on the international climate change negotiations. In an equitable world, everyone should have access to affordable and clean energy. The emissions cut, closely linked with energy use, have become an issue of discord in international relations. The developed countries, which are threatened by the prospect of climate refugees swamping their populations, are pressurising the developing countries like India and China to reduce their emissions and assume targets for emission reduction. The developing countries, on the other hand, are insisting that the West should take responsibility for the present levels of CO_2 emissions in the world on the basis of the principle of "common but differentiated responsibility" and pay for mitigation and adaptation measures. Reduced consumption of energy by the developing countries will only increase poverty and hinder development which could further destabilise the world and worsen environmental problems. The developing countries like India have projected the climate change issue as one of equity and justice. The developed countries, on the other hand, see the emerging economies like India and China as major emitters whose emissions will surpass those of the developed countries in the near future.

One of the potential causes for instability in the world could be the likelihood of large-scale migrations due to climate change. Migration is already a major international phenomenon today. Migrations take place due to a variety of reasons. Not all migrations are harmful. Migrants send back home nearly $232 billion to their countries of origin. However, disruptive migrations which take place due to conflicts, famines, disease, wars, disasters, etc., can lead to serious security problems. On several occasions in the past the US, Australia, Italy, France, Spain, Portugal, Malaysia and many other countries have used their security forces, including the military, to prevent illegal migrants from entering their territories. Russia regards illegal migration as a national security threat. Africa is considered to be a potential source of large-scale migrations. The important point here is that while migration is a natural state of affairs in the world today—there were 191 million migrants in the world in 2005—the trends of disruptive migration could intensify due to climate change. The US is becoming sensitive to migration. It is facing large-scale illegal migration from Mexico and the Caribbean Islands. Thousands of people migrated permanently out of Louisiana in the wake of hurricane Katrina in 2005. While it is difficult to precisely pin down climate change as a reason for migration because migrations take place due to a variety of inter-related reasons, it is estimated that there would be 50 million migrants by 2010 due to environmental factors alone.

As can be seen from the brief discussion above, climate change may lead to instability in countries, particularly the poorer ones, who may not be able to cope with the adverse impact of climate change. The lack of capacity to adapt to climate change means that these countries will be pushed into a spiral of scarcity, conflict and violence. The resulting instability could affect the entire international system.

Securitisation of Climate Change

Climate change presents a non-traditional security threat. This is widely accepted in security studies literature and also in international security discourses on non-traditional security threats. Climate change, terrorism, pandemics, financial institutions, etc., are diffused, have cross-border character and affect human security directly. Thus, security implications of climate change, whether at the level of international security, national security or human security, cannot be ignored. However, 'securitisation' of climate change, as reflected in the approach of a few Western countries, is a different matter. 'Securitisation' of climate change appears to be agenda-driven. The West's agenda appears to be to use the security dimension of climate change to force the developing world to fall in line on climate change negotiations and pressurise them on governance-related issues. It is important to keep the distinction between the security dimension and securitisation of climate change.

The international community is finding it difficult to come to grips with climate change-related issues. There is an increasing trend in the West to treat climate change as a security issue. The North and the South are deeply divided over emission cuts and mitigation measures. A key issue is: who is responsible for climate change? The developing countries, with considerable justification, blame the developed countries for the present levels of CO_2 emissions. Under pressure, the developed countries have assumed modest obligations for emissions cut under the Kyoto Protocol but they are pressurising the developing countries, particularly India, China, Brazil, etc., to restrain their own emissions by assuming quotas for emission cuts. It is not clear whether the modest obligations undertaken by the developed countries to reduce CO_2 emissions will be fulfilled by them. Several developed countries have missed their emission targets. They are not prepared to change their wasteful lifestyles while they continue to pressurise the developing countries to reduce their emissions which have adverse impact on development. The ranks of the developing countries have also been split. The small island countries, faced with the threat of extinction within the century are keen that emissions, regardless of where they come from, should be curtailed so that the adverse effects of climate change can be arrested.

The developed countries have adopted a new tactic to pressurise the developing countries—that of linking climate change with security issues. There have been a number of studies in recent years which have linked climate change with national and international security issues. On April 17, 2007 the UK, as the Chairman of the UN Security Council (UNSC), took the initiative to organise an open debate at the UNSC to discuss the security implications of climate change. About 50 countries took part in the debate. The message from the West was clear: climate change is an issue that has implications for international security. Several developing countries differed. They held that climate change was essentially a socio-economic issue. The UNSC was not the right forum to discuss the issue as a security issue. Climate change is not a Chapter VI and Chapter VII issue. So, it does not fall under the purview of the UNSC. The issue should be discussed at the UNFCC, the Commission for Sustainable Development and Economic and Social Council of the UN. The West should not securitise the issue. Instead it should focus on how to help the developing countries with finances and technologies etc., to strengthen their capacity to adapt to climate change. However, several countries in Africa and the Pacific went along with the Western thesis that climate change had security implications. The UNSC did not adopt a resolution. But the writing on the wall was clear: the West is trying to securitise the debate on climate change. Securitisation of the issue could lead to sanctions and other such measures which are inadequate to address climate change.

Although securitisation of climate change is counter-productive, as this only increases the hiatus between the developed and the developing countries, there is no denying that climate change can have security implications. Climate change is regarded as a non-traditional security threat. That is quite obvious even if the available analysis of the impact of climate change is patchy, incomplete and often motivated by the desire of the developed countries to shape a new security agenda.

Admittedly, the theoretical framework for exploring the link between climate change and its security implications is as yet not robust enough. The predictive power of the models adopted is low. Many of the scenarios, linking wars with scarcity, seem to be exaggerated. But, it cannot be denied that human security is intimately tied up with the health of the environment. Any degradation of the environment, whether natural or man-made, creates stress. Such stresses will be aggravated by climate change. At the same time, it must be accepted that most of the dire scenarios projected above may not come to pass or they may not manifest in so severe a fashion, if suitable preventive action is taken to stabilise carbon emissions. Securitising climate change can serve the purpose of alerting the international community to take action now before it is too late. At the same time, it may create panic and concern which may lead the States, particularly the developed ones, to create artificial barriers and fortresses which can adversely impact international cooperation. Presenting climate change solely as a threat to national security and international peace while ignoring the socio-economic and other dimensions, may compel many countries to consider military solutions to the problem. That would be highly counter-productive.[10]

A host of recent studies in the US, the EU, Australia and other developed countries have focused on climate change as a national security issue. Studies in the US have listed how the US military bases around the world will be affected by climate-related factors and how the US military should be prepared to face the consequences of a general rise in instability and terrorism around the world due to climate change. Climate could affect countries with whom the US has alliance relations. A recent assessment by the National Intelligence Council—with input from all 16 US intelligence agencies—treats climate change as a security threat. The intelligence assessment found Sub-Saharan Africa, the Middle East and Central and Southeast Asia to be the most vulnerable to warming-related drought, flooding, extreme weather and hunger.[11]

The UK has directly linked climate change with war and conflict. Britain used its presidency of the UN Security Council to lead its first debate on climate change and conflict. "What makes wars start?" asked Mrs. Margaret Beckett, the chairperson of the UNSC in April 2007. "Fights over water. Changing patterns of rainfall. Fights over food production,

---

10. For the views of different countries expressed at the UNSC open debate "On Impact of Climate Change on Peace, Security", see Security Council SC/9000, dated 17 April 2007.
11. "Intelligence Agencies: Climate Threatens Security", 25 June 2008 at *http://www.npr.org/templates/story.php¿storyId=91878515_*

land use. There are few greater potential threats... to peace and security itself."[12] Such claims tantamount to over-reaction and scaremongering.

Although securitisation of climate change is counter-productive, there is no denying that climate change can have security implications. That is quite obvious even if the available analysis of the impact of climate change is patchy, incomplete and often motivated by the desire of the developed countries to shape a new security agenda. The important issue is: what should be done to address climate change and its impacts, socio-economic or any other? The German Advisory Council on Global Change (WBGU), Berlin, produced a report in May 2007 which focused on an action plan to deal with the adverse impact of climate change. The WBGU report considered climate change induced inter-state wars as unlikely but new "conflict constellations"—degradation of water resources, decline in food production, increase in storm and flood disasters and environmentally induced migrations—as likely. It identified several regional hotspots in North Africa, Sahel zone, Southern Africa, Central Asia, India, Pakistan, Bangladesh, China, the Caribbean and the Gulf of Mexico, the Andean region and Amazonia. One of the recommendations it makes is preventing conflict by supporting adaptation strategies for developing countries.[13] Such an approach to climate change is balanced and sensible.

A study for the Royal United Services Institution (RUSI) done by Nick Maby (2008) examines the possible international "security responses" in a world which is being transformed by climate change. He feels that only a transition to low carbon economy can reduce tensions in the world. The world has sufficient financial and technological resources to deliver a low carbon economy but not enough has been done. One of his conclusions is that, "Failure to adequately address the interests of the poorest countries and people in a global climate change agreement will seriously undermine the legitimacy and effectiveness of the international system."[14] He also suggests involvement of the security community in discussions about the mitigation challenge.[15]

---

12. *op. cit.*, no.10.
13. German Advisory Council on Global Change, "World in Transition: Climate Change as Security Risk", Summary for Policymakers, available at http://*www.wbgu.de*
14. Maby, Nick (2008). "Delivering Climate Security: International Security Response to a Climate Changed World", *Whitehall Paper* 69. Oxford, UK: The Royal United Services Institution, Routledge. p.95.
15. Ibid.: 53.

Climate Change and National Security: The Indian Context

The link between climate change and national security works both directly and indirectly. In India, climate change is expected to have a deep impact on food, energy, water, demography, health and economy. It could aggravate numerous existing conflicts. At the same time, it could generate new stresses, tensions and conflicts. Climate change could be an issue in India's bilateral relations with its neighbours. Climate change obligations could also restrict India's energy choices and put limits on economic growth.

The South Asian region is highly vulnerable to the adverse impact of climate change. According to the various studies, discussed later in this report, the pattern of monsoons in South Asia will become erratic. The overall precipitation may increase but there will be considerable regional variations. The Himalayan glaciers may also melt significantly in the next 40 to 50 years. More floods, droughts, desertification and extreme weather conditions are predicted for the region. The rise in sea level will have a serious impact on the coastal regions of Bangladesh and India. The Maldives and the island regions of India will be submerged to a large extent.

Such geophysical changes will lead to soil erosion and increase in vector-borne diseases. The variation in temperature and monsoon, the soil erosion due to floods and rise in sea level will create food and water scarcity. The adverse impact on human security would play on national security as well. There could be more hunger, more law and order problems, overcrowding of cities as a result of food and water scarcity. Since many of the countries in South Asia have weak governance structures, inadequate public health systems and low adaptation capacity, there would be greater strain on governments' resources. Climate change could create chronic economic problems including unemployment. Unemployed youth could take to militancy, terrorism and organised crime. It could create fresh conflicts due to environmental reasons and also aggravate the existing conflicts. Environmental distress due to climate change will create pressure on populations to migrate to safer locations in search of livelihood.

India could also face bilateral problems with neighbouring countries due to climate change. For instance, a large number of environmental refugees could come to India from Bangladesh, thereby altering the demographic balance in the Indian states. Migrations could also take place from Sri Lanka, the Maldives and Nepal. Water sharing issues could arise between India and China, India and Bangladesh, India and Pakistan.

The frequency and intensity of natural disasters including hurricanes, floods and famines may increase. This will put greater strain on the armed forces in disaster management. The border security forces will have to contend with increased migration flows from the neighbouring countries. The armed forces may also be compelled to change their war-fighting doctrines and operations in the light of climate change. They may be required to undertake new missions as those that happened during the Tsunami in 2004, connected with disaster management domestically and internationally.

These issues are discussed in the succeeding chapters. A note of caution is in order. The above list is merely indicative of what is likely to happen. Precise quantitative predictions are not possible. The likely impacts are based on models whose power to predict is often limited. (A lot more research, focusing on India-specific issues needs to be done to understand the implications of climate change for India.)

The probability of such events taking place is difficult to calculate. Moreover, the impact of climate change would not be sudden. It would unfold gradually over several decades, thereby providing time for adaptation. Therefore, it would not be proper to sensationalise the national security impact of climate change for India. At the same time, it would be counter-productive not to be aware of the lurking danger.

Since there is still considerable uncertainty about climate change, its impact and its link with national security, it may be useful for India to adopt a risk-based approach. It may be better to put in place "no regrets" measures which will stand India in good stead.

In order to deal with these issues, India will have to think through its strategies carefully. On the one hand, it will have to resist the international pressure on it to take binding objections with regard to CO_2 emissions as this will seriously affect the prospects of economic growth, on the other hand, it will have to undertake suitable adaptive measures to ensure that its economic growth is based on sound principles of energy efficiency resource conservation. Since India is being labelled as a large emitter, it will have to craft its negotiating position to safeguard its national interests without being isolated. The present report discusses some of these issues in the subsequent chapters. An effort has been made to take cognisance of the scientific and technical studies which pertain to the impact of climate change on the South Asian region, including India.

2 Key Vulnerabilities Due to Climate Change

Murari Lal and Sunil Chauhan

Introduction

This chapter analyses the observed changes in climate and the projected climate change in the Indian context and the potential risks and sectoral impacts India could face in the next few decades. The analyses are based on the studies, models and projections carried out by the Government of India (GoI) as part of India's Initial National Communication (NATCOM) to the United Nations Framework Convention on Climate Change (UNFCCC), the GoI/UK's Department for Environment, Food and Rural Affairs (DEFRA) 2005 programme and other studies. The findings reported in the technical document of the Indian National Action Plan on Climate Change (NAPCC) released in June 2008 by the Prime Minister's Office have also been referred to in preparation of this chapter.

Current Understanding of the State of the Climate System

Today, there is concern in all countries about the national and global impacts of climate change and variability. The Fourth Assessment Report (AR4) of the Intergovernmental Panel on Climate Change in 2007 demonstrated convincingly that most of the global warming over the last 50 years is attributable to human activities. The current state of our understanding of the climate system is as under:

- Since the Industrial Revolution, the concentrations of greenhouse gases in the earth's atmosphere have increased due to the combustion of fossil fuels (such as coal, oil and gas), industrial, energy-related and land use activities.
- The atmospheric abundances of almost all greenhouse gases have now reached their highest values in recorded history. The pre-

industrial concentration of CO_2 was 280 parts per million (ppm). In 2005, in Mauna Loa, Hawaii, one of the most pristine parts of the world, this had reached 381 ppm. In 2006, measurements in Svalbard, in the high Arctic region, had recorded 390 ppm—a full 10 ppm above the global average.

- Global CO_2 emissions from the burning of fossil fuels stood at a record 8.38 gigatons of carbon (GtC) in 2006, 20 per cent above the level in 2000. Emissions grew 3.1 per cent a year between 2000 and 2006, more than twice the rate of growth during the 1990s.
- A mean global warming of about 0.8°C has been observed since the late 19th century. Recent years have been exceptionally warm, with a larger increase in minimum than in maximum temperatures, possibly related, among others, to an increase of cloud cover. Globally, the Earth's climate is warmer today than at any time during the last 140 years.
- Over the Indian subcontinent, an area-average increasing trend of close to 0.57°C/100 years has been recorded in the 20th century (Figure 2.1). A warming trend has been observed along the west coast in central India, the interior peninsula and northeastern India. However, cooling trends have been observed in northwest India and parts of south India.[1,2]
- At the all-India level, the observed monsoon rainfall has not shown any significant trend during the 20th century (Figure 2.2), and the variations in space and time have been rather random in nature. A trend of increasing monsoon seasonal rainfall has been observed along the west coast, northern Andhra Pradesh and northwest India (+10 per cent to +12 per cent of the normal over the last 100 years), while a trend of decreasing monsoon seasonal rainfall has been observed over east Madhya Pradesh, northeast India and some parts of Gujarat and Kerala (-6 per cent to -8 per cent of the normal over the last 100 years).[3, 4]

---

1. GoI (2008). *National Action Plan on Climate Change (NAPCC)*. Prime Minister's Council on Climate Change. Available at *www.pmindia.nic.in*
2. GoI (2004). *India's Initial National Communication (NATCOM) to the United Nations Framework Convention on Climate Change (UNFCCC)*. Ministry of Environment and Forests (MoEF).
3. GoI, 2008 *op. cit.*
4. GoI, 2004 *op. cit.*, p.vi-vii, 61-62.

- In India, the frequency of extreme weather events—for example, heat waves, droughts and floods—has increased over the past two decades. The state of Orissa has been reeling under contrasting extreme weather conditions for more than a decade: from heat waves to cyclones and from droughts to floods. The death toll due to heat waves from March to May in Rajasthan, Punjab, Gujarat, Orissa and Bihar is on the rise in recent years.
- The Himalayas have nearly 1,500 glaciers and it is estimated that these cover an area of about 34,000 square km. These glaciers provide the snow and the glacial-melt waters keep our rivers perennial throughout the year. Almost 67 per cent of the glaciers in the Himalayan mountain ranges have retreated in the past decade. Surface air temperatures in the high Himalayas have risen by 1.5°C since the mid-1970s. Available records suggest that the Gangotri glacier is retreating about 30 metres per year.
- Oceans and the atmosphere are changing at a rate faster than previously experienced. Most of the increased heat of the earth during recent decades has gone into the ocean. The increased energy of the ocean atmosphere system is driving an increase in more extreme weather events. Storm intensity is about five times greater than expected, and cyclone intensity has grown as well. Furthermore, sea level is rising almost by a factor of two, faster than it did during the half century prior to 1990. Moreover, the El Niño episodes (unusual warming of the Equatorial Pacific Ocean) since the mid-1970s have been relatively more frequent than the opposite La Niña episodes.
- The observed rate of rise in sea level along the Indian coastline have been estimated to be between 1.06 to 1.75 mm per year, based on data and records of the coastal tide gauges in the north Indian Ocean for more than 40 years.[5] These rates are consistent with the 1-2 mm per year global sea level rise estimates of IPCC.[6] The sea level rise of 0.4 to 2.0 mm has been recorded along the Gulf of Kutch while the rise along the coast of West Bengal is the highest. Along the Karnataka coast, there is a relative decrease in the sea level.[7]

---

5. Unnikrishnan A.S. and D. Shankar (2007). "Are sea-level rise coast of the trends along the north Indian ocean consistent with global estimates?", *Global and Planetary Change* 57, 301: 15.
6. GoI, 2008 *op. cit.*
7. GoI, 2004 *op. cit.* p.ix.

- Frequent inundation of low-lying areas, drowning of coastal marshes and wetlands, enhanced erosion of beaches, more flooding and increased salinity of rivers, bays and aquifers in the coastal regions of India have occurred in recent years. Today, low lying areas and coastal megacities (e.g., Mumbai, Kolkata and Chennai, etc.) are subject to increased erosion and loss of coastal protection from natural ecosystems such as coral reefs and wetlands. Rising sea temperatures have been responsible for recent large-scale bleaching and mortality along a long Indian coastline. Sea surface temperatures throughout the tropics have shown dramatic increases over the last two decades; as much as half a degree per decade. This is 10 times what we are observing globally. As a result, the concern for coral reefs is how much of this increase will continue over the ensuing decades.
- The biodiversity of coral reefs includes sea grasses, corals, several invertebrate groups, fish, amphibians, birds nesting on reefs and mammals. The reefs of the Andaman and Nicobar Islands have the highest recorded diversity with 203 coral species, 120 algal species and 70 sponges in addition to fish, sea turtles, dugongs and dolphins. The first signs of sea-level increase and consequent submergence of coastlands are already showing up in the 10,000 km stretch of the Sundarbans, the world's largest mangrove wetland, in the Ganges-Brahmaputra delta in eastern India. There are reports from this Islet region of a number of small islands disappearing underwater in recent years, causing their inhabitants to become destitute.
- Combined with coastal population growth trends, India is threatened with a growing tide of environmental migrants either permanently or seasonally as millions would be displaced due to an increased intensity and frequency of storms.
- The fishery resources of India are being depleted by increased fishing, excessive use of pesticides and industrial pollution. The increase in marine culture products and decline in the marine fishery output are the current trends in commercial fishery activity. Life supporting food production in many coastal areas of South Asia is changing and threatened, including the location of fish populations, and food security is at risk.

- China has grown faster than India but has also consumed over 11 times the fossil fuels since 2002. India on the other hand, has been delivering an 8 per cent GDP growth with only 3.7 per cent growth in its energy consumption. Nonetheless, India is the 5th largest—and growing-emitter of GHGs today—at a time when the window for remedial action is reducing. It may be worth noting here that the global growth rate of CO_2 between the year 2000 and 2006 has been close to 3.3 per cent per year as against the highest growth rate of 2.4 per cent projected by IPCC under the A1FI emission scenario in 2001 (Figure 2.3).

Figure 2.1

Observed Trends in All India Mean Surface Air Temperature during the 20th Century

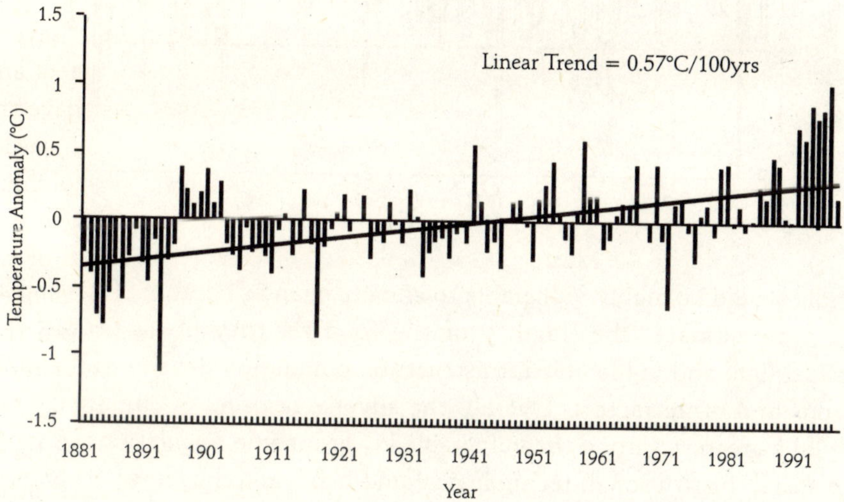

The IPCC AR4 in 2007 demonstrated convincingly that most of the global warming over the last 50 years is attributable to human activities and the projected temperature increase by the end of this century would be between 1.5° and 6°C. Assuming that the pace of climate change over the next 50 years will match the IPCC's 2007 projections, the world's average surface air temperature will increase beyond the critical threshold of 2°C. New studies also suggest that today, the Arctic Ocean's ice cover is 40 per cent thinner than 20 to 40 years ago and will contribute significantly (in addition to the thermal expansion) to the global rise in sea levels. Studies further suggest that global warming increases the risk of extreme weather, including floods, droughts and heat waves. The melting of permafrost will release more carbon dioxide and methane into the atmosphere and so accelerate the process of global warming.

Figure 2.2

Observed Anomalies in All India Mean Summer Monsoon Rainfall during the 20th Century

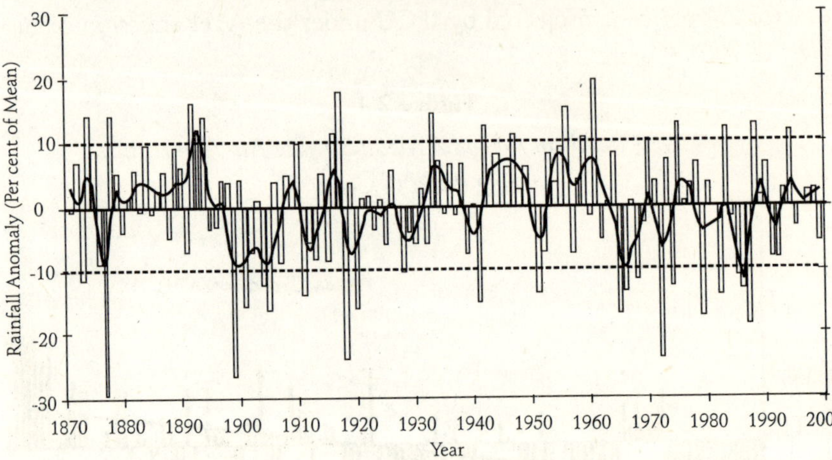

India would be highly vulnerable to climate change, because of its already extreme climate, the fragility of the food security of its people, the inadequate and vulnerable infrastructure, population density in marginal lands and other factors. Overall, the adverse impacts of climate change will be greatest among the vulnerable lower income populations. A range of social, institutional, technological and behavioural adaptation options would need to be practiced to lessen the impacts of climate change in India. In order to access the key vulnerabilities on various sectors as a consequence of climate change in India, a confident regional and local climate change scenario is required along with a plausible state of resource availability under varying trends of population pressure, economic development, land-use changes and energy demand and supply gaps. The following parts of this chapter present an overview of our current state of knowledge on the most plausible climate changes in India.

Figure 2.3

Observed Growth Rates in Global CO_2 Emissions as Compared to IPCC's SRES Marker Emission Scenarios

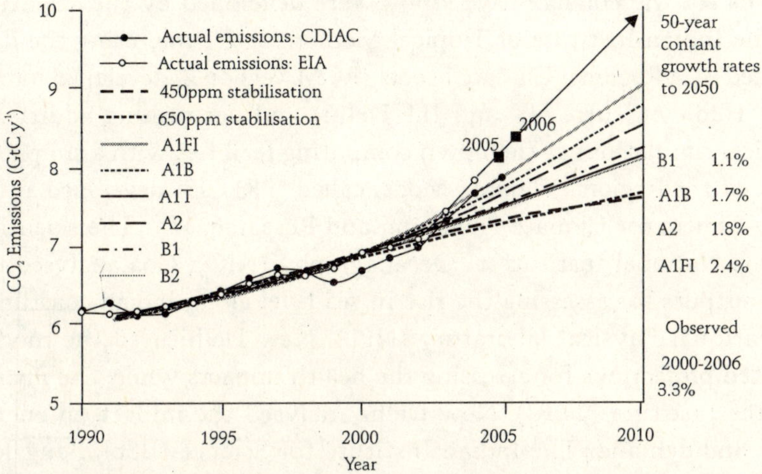

Climate Change Scenario in India

The climate change scenario for Indian subcontinent has been assessed from data generated from coupled atmosphere ocean global climate models (AOGCMs) by researchers from the Indian Institute of Technology, Delhi, during the past two decades. These scenarios were recently updated from more recent simulations of as many as eight AOGCMs and consolidated as part of the NATCOM project executed by the Ministry of Environment and Forests (MoEF). The base line simulations in these model experiments were corresponding to the 1961-1990 period and the enhanced forcing simulations for two emission scenarios namely, the Special Report on Emission Scenarios (SRES) A2 (medium-high) and SRES B2 (medium-low) scenarios[8] extended to the end of 21st century (some earlier model simulations also used the outdated IS92a emission scenarios).

8. *IS92a Scenario of GHG increase:* In this the GHG forcing is increased gradually to represent the observed changes in forcing due to all the GHG from 1860 to 1990. For the future time period 1990-2099, the forcing is increased at a compounded rate of 1 per cent per year (relative to 1990 values), representing the IS92a emissions scenario.

   A2 Scenario of SRES (A2): A2 scenario falls in the category of 'Medium-High' emissions. The cumulative global carbon emissions between 2000 and 2100 for this scenario is taken to be 1862 GtC (1GtC = 1 giga or Bt of Carbon;1 tonne of Carbon = 3.67 tonnes of CO_2).

   B2 Scenario of SRES (B2): B2 scenario falls in the category of 'Medium-Low' emissions.

In addition to the NATCOM, a three-year collaborative research project on "Investigating the Impacts of Climate Change in India" was also carried out by the MoEF, India and DEFRA, UK. The climate change scenarios for the collaborative study were developed by the scientists from the Indian Institute of Tropical Meteorology, Pune, using the data generated by a Regional Climate Model (RCM Version 2, developed jointly by the Hadley Centre UK and IIT Delhi) and performing additional numerical simulations at their own computing facilities with a simplified version of the Regional Climate Model called "PRECIS" developed at the Hadley Centre for Climate Prediction and Research, UK. The scientists from the National Institute of Oceanography (NIO), Goa analysed the model outputs for assessing the rise in sea level along Indian coastlines. The National Physical laboratory (NPL), New Delhi used the model-generated projections for assessing the health impacts while The Energy Resource Institute (TERI), New Delhi analysed the impact on energy supply and demand. The Indian Institute for Science (IISc), Bangalore applied the model-generated data for impacts on India's forests and ecosystems while the Indian Agricultural Research Institute (IARI), New Delhi examined the implications of projected climate on food production in India. The Indian Institute of Management (IIM), Ahmedabad assessed the impacts on population and economic growth in India. This research documented in the GoI/DEFRA 2005 Report provided projections of likely changes in surface air temperature and rainfall along with possible perturbations in monsoon characteristics and extreme events over the Indian subcontinent. The detailed findings from NATCOM and GoI/DEFRA studies by individual Indian scientists have subsequently been published as a special issue of the *Current Science* (February 2006).[9]

The subsequent paragraphs present the climate change scenario in India using the results reported in the above mentioned studies, supplemented as necessary with more detailed and confident model-based data sources through recent studies. Important results and findings of some of these studies also form the basis of the technical document in NAPCC released in June 2008 by the Prime Minister's Office. More recent

---

9. Special Section (2006). "Climate Change and India." In: *Current Science* 90(3): 314-395. 10 February.

and perhaps more confident projections are under way for the Second Indian National Communication (SNC) to the UNFCCC, results of which however, are not available as of now.

Projections of Surface Air Temperature Changes

As far as the temperature trends are concerned, all the models suggest positive trends over India, indicating widespread warming into the future. The spatial patterns of annual mean temperature changes in different model simulations indicate that the warming is more pronounced over the northern parts of India. Also, the warming is more during winter and post-monsoon months compared to the rest of the year. Interestingly, this is a conspicuous feature of the observed temperature trends from the instrumental data analyses over India.

The analysis of future projections of surface warming over India indicates that over the Indian subcontinent, the annual mean area-averaged surface warming is likely to be between 2.0° and 3.0°C and 3.5° and 5.5°C by the middle and end of the 21st century respectively [1-4] [time evolution of annual and seasonal climate change projections

Table 2.1

Climate Change Projections for Indian Subcontinent with Respect to Time under the SRES Marker Emission Scenarios*

		Temperature Change (°C)				*Rainfall Change (%)*			
Scenarios -		*A1*	*A2*	*B1*	*B2*	*A1*	*A2*	*B1*	*B2*
2020s	Annual	1.18	1.00	1.32	1.41	2.29	2.16	4.15	5.97
	Winter	1.19	1.08	1.37	1.54	0.39	-1.95	4.36	3.64
	Monsoon	1.04	0.87	1.12	1.17	1.81	2.37	3.83	5.10
2050s	Annual	2.87	2.63	2.23	2.73	9.34	5.36	6.86	7.18
	Winter	3.18	2.83	2.54	3.00	3.22	-9.22	3.82	3.29
	Monsoon	2.37	2.23	1.81	2.25	10.52	7.18	7.20	8.03
2080s	Annual	5.09	5.55	3.53	4.16	9.90	9.07	7.48	7.62
	Winter	5.88	6.31	4.14	4.78	-19.97	-24.83	-4.50	-10.36
	Monsoon	4.23	4.62	2.91	3.47	14.96	15.18	11.12	10.10

Note: * Based on analysis of model data generated by the CCSR/NIES (Japan) Model Experiments; Area averaged for land regions only.

for various trajectories (SRES emission scenarios) of global emissions of greenhouse gases in the earth's atmosphere are detailed in Table 2.1]. The warming would be more pronounced in winter than during the monsoon season [5-7].

Recent press reports in India have quoted a warming of over 4°C over India by the mid-21st century predicted by the Providing Regional Climate for Impact Studies (PRECIS) model at the Indian Institute of Tropical Meteorology, Pune. However, it would not be incorrect to cast strong reservations about the confidence level of these predictions, as the current model being deployed by IITM, Pune is a simplified version of the highly complex UK Hadley Centre model and many components of the physical and dynamical forcings and feedbacks are missing herein. The monsoon circulation and the associated rainfall over the Indian subcontinent are strongly modulated by the forcing as is evident from Figure 2.4 below. These constraints limit the validity of such projections and a prudent

Figure 2.4

The Contrasting Effects of Radiative Forcing due to Greenhouse Gases and the Aerosols on the Monthly Mean All India Mean Rainfall as Simulated by Climate Models

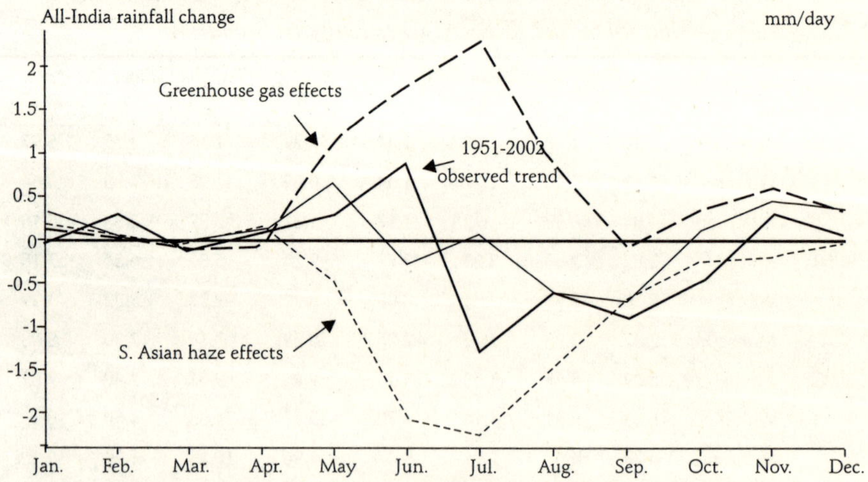

approach would be to undertake more stringent validation of a group of more complex high-resolution climate models available across the world and then perform multiple simulations with perturbed forcings before passing on the findings to the media, that may create unwarranted panic.

Rainfall

Analysis of future projections of rainfall over India indicates that India may experience between 5 and 25 per cent decline in winter rainfall by the end of the 21st century while only about a 10 per cent increase in area-averaged monsoon rainfall is likely over the Indian subcontinent [8, 9]. The decline in winter rainfall over India is likely to be significant and may lead to droughts during the dry summer months [3, 4]. Most model projections, indeed, indicate enhanced precipitation during the monsoon season by the end of the 21st century, which is particularly significant over the northwestern parts of India. Also, no significant change in the monsoon rainfall over most of peninsular India is projected. However, the magnitudes of projected change differ considerably from one model to thc other. PRECIS projections exhibit a decline in monsoon rainfall over northwestern India which makes the model projections questionable. The PRECIS projections also suggest a year round increase in all-India mean rainfall (Figure 2.5) and a 10 per cent increase in monsoon rain intensity by the end of the 21st century under A2 scenario of the IPCC.

A review of the projections and different models also reveals the year-to-year variability in rainfall in central India during the monsoon season is not likely to change significantly in the future. [6, 8] However, the date of the onset of the summer monsoon over central India would become more variable [10, 11]. More intense spells of rainfall are also projected in a warmer atmosphere, increasing the probability of extreme rainfall and hence flash floods in parts of India [4, 12-15].

Figure 2.5

Projected Changes in Monthly Mean Surface Air Temperature and Precipitation Averaged over India by the End of the 21st Century for A2 and B2 Scenarios

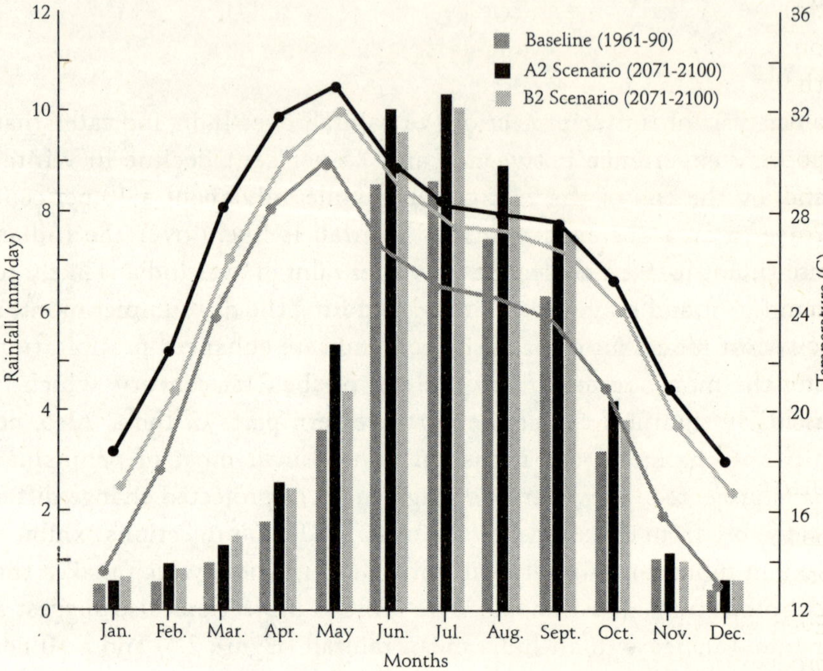

Source: Kumar, Rupa *et al.*, 2006.

Projected Rise in Sea Level

Precise model-based projections of the rise in sea level along the Indian coastline for the 21st century has not been reported in any published study so far. However, estimates for the likely rate of the rise in sea level through observed data correlated with the IPPC, predicted a sea level rise rate of 10-25 cm per 100 years are available in the NATCOM 2004 report.

- Model projections using the HadRM2 were made in the GoI/ DEFRA 2005 assessment to analyse the impact of climate change on extreme sea level rise, *viz.*, 'storm surges' (see Chapter 3).
- Recent assessments by Indian scientists analyse trends in the rise of sea levels through observed data of past 40 years along the

Indian coast and estimate sea level rise between 1.06 to 1.75 mm per year[10] consistent with the IPCC rate of 1-2 mm per year.[11] The report identifies the northern areas along the Indian coast as more vulnerable to rise in sea levels.

- Analysis on likely rise in sea level along the Indian coastline based on model data from available coupled AOGCMs suggests that the maximum increase along the Indian coastline is likely to be along the Gujarat and Maharashtra coasts. The security of coastal populations in India is on a risk trajectory due to rise in sea levels and also due to greater intensity and frequency of storms [7, 17, 26]. A possible increase in cyclone intensity of 10-20 per cent for a rise in sea surface temperature of 2° to 4°C relative to the threshold temperature of 28°C, is very likely in Indian seas [17-25]. Hence, low lying areas and coastal megacities would be subject to increased erosion and loss of coastal protection from natural ecosystems such as coral reefs and wetlands [7,17].

Key Vulnerabilities of Climate Change for India

The key vulnerabilities in India in terms of water and food insecurity and more extreme weather conditions could be identified as under:

- Even a 1° to 2°C temperature increase would lower agricultural productivity. Decreased yields in non-irrigated wheat and in rice are very likely, and a loss in farm level net revenue between 9 per cent and 25 per cent would take place for a temperature increase of 2°-3.5°C [27-36].
- Shortage of drinking and irrigation water in drought-affected areas of India would cause food shortages, and in turn, human malnutrition [37-47].
- In areas of frequent flooding, much of the freshwater supplies would become contaminated with human waste and bacteria, further decreasing the amount of drinking water available and creating sanitation issues [41-53].

---

10. Unnikrishnan and Shankar, 2007 *op. cit.*
11. GoI, 2008 *op. cit.*

- Increased frequency of El Niño events likely in a warmer atmosphere, [54, 55] could lead to measurable declines in fish larvae in coastal waters of South Asia [7, 10]. Climate change will also alter marine ecosystems and resources, including fish populations, and our abilities to depend on the location and abundance of these populations [7]. This could have potentially severe effects for some coastal states in India (Kerala, Orissa and West Bengal), particularly those dependent on fish as a source of protein intake [10].
- Mangroves are critical biosystems which grow in inter-tidal regions and act as a primary nursery area for a number of commercially important shrimp, snappers, crab, penaeidae prawns, sea perch, catfish and clams. They are also the breeding ground for other species like angiospermic flora, flagellates, phytoplankton, seaweeds, sea grasses, sponges, corals and benthic algae which sustain the oceanic food chain [7,10,17]. Global warming and rising sea levels will significantly shrink mangrove areas along the Indian coastline, resulting in devastating effects on the marine and coastal food chain and ecosystem.
- The reduction of marine biodiversity may be associated with exponential reductions of ecosystem functions: 20-25 per cent species loss can cause a reduction of 50-80 per cent of ecosystem functions, threatening our life supporting services [17].

Impact on Agriculture and Food Production

Agriculture and food production in India is sensitive to climate changes within a season, especially surface air temperatures and rainfall. The NATCOM and GoI-DEFRA Assessment Reports indicate expected loss in *rabi* crop is likely to be greater. Every 1°C rise in temperature would reduce wheat production by 4-5 million tonnes. A loss of 10-40 per cent in crop production has been projected by 2100.[12] The findings also suggest lower yields from dairy cattle and decline in fish breeding, migration and harvests.

Assessments based on review of published work till date suggests that the water scarcity in India will dramatically increase vulnerability and

---

12. GoI, 2008 *op.cit.*, p.16.

threaten food supplies in India [64]. The climate change-induced impacts on agricultural productivity by 2050 would be:

- Climate change will lead to generally negative responses and grain yields are likely to register an overall decline under the combined influence of CO_2 fertilisation and temperature rise [65]. The farming community may opt to switch over to crop varieties which are less demanding on water resources. More attention would need to be focused on horticulture productivity in peri-urban areas to meet increasing demand.
- Climate change may have a dramatic impact on fish production, reducing the supply of fish meal and fish oils. The market structure for fish meal is a key factor in determining whether increasing aquaculture production can affect fish meal prices. Increasing sea surface temperature has the potential to increase the intensity and frequency of disease outbreaks, exerting negative effects on marine aquaculture.
- A shift from meat and poultry products including fisheries to non-meat products is possible to avert the health-related problems and as a proposed adjustment to declining fisheries productivity and scarcity of animal feeds.
- Impacts of climate change on agricultural productivity and prices are estimated to result in trade deficit and income losses in India [66].

Impact on Water Resources

The availability of water is likely to be altered in significant ways due to climate change. About 67 per cent of the nearly 34,000 sq km of Himalayan glaciers are receding and in the long run as the ice diminishes, glacial runoffs in summer and river flows will also go down, leading to severe water shortages in the region [63]. The available data on Himalayan glaciers indicates that while recession of some glaciers has occurred in some Himalayan regions in recent years, the trend is not consistent across the entire mountain chain.[13] However, the glaciers form the main source of water for the perennial rivers such as the Indus, the Ganga and the Brahmaputra. Glacial melt will impact their long-term lean season flows

13. GoI, 2008 *op. cit.*

with adverse impact on the economy in terms of water availability and hydropower generation.

During one monsoon season in India, river basins can face both floods and droughts. With climate change likely to cause more intense and longer floods and droughts, the whipsaw effect of being pulled in one and then the other direction may be far more devastating than simply dealing with a serious flood. The NATCOM project used the Soil and Water Assessment Tool (SWAT) water balance model for hydrologic modelling of different river basins in the country, in combination with the outputs of the HadRM2 regional climate model to predict the impact of climate change on the water resources in India.

Assessments based on review of published work till date suggests that the climate change-induced impacts on water resources in India by the year 2050 would be:

- Flood magnitude would episodically increase as a consequence of increased frequency of heavy precipitation events during the monsoon season, which can increase runoff but curtail ground water recharge in some floodplains. Significant degradation is also expected in water quality.
- Water demand for irrigation would increase necessitating improvements in irrigation efficiency.
- Sustainable consumption of fresh water needs to be practiced to conserve water resources available for maintaining biodiversity and ecological functions and services as strategies for responding to the impact of climate change on water resources.

Impact on Forests

Rising temperatures and a retreating snow-line in Himachal Pradesh have fatally affected its once-legendary apple industry and crippled local economies in recent years. Up to 50 per cent of the country's flora and fauna could be threatened, with at least a quarter of our biodiversity lost. Nearly 68-77 per cent of Indian forests are likely to undergo a sea change to another forest type by 2085 [67]. Over half of the vegetation is likely to find itself less optimally adapted to its existing location, making it vulnerable to adverse climatic conditions and to biotic stresses. The actual negative impact may be even more because of a 'domino effect' on other species due to the inter-dependency of plant-animal-microbe communities. Assessments by Indian scientists[14,15] based on the climate projections of the Regional Climate Model of the Hadley Centre, UK (HadRM3) using the A2 and B2 scenarios and the BIOME4 vegetation response model, indicate:

- 77 per cent and 68 per cent of the forested grids in India are likely to experience a shift in forest types under A2 and B2 scenarios 77 per cent respectively by end of the century (2085).
- Indications are a shift towards wetter forest types in the northeastern region and drier forests types in the northwestern regions.
- Consequent changes in forests produce and in turn on the livelihoods based on these products. Associated biodiversity is likely to be adversely affected.

One-tenth of the world's known species of higher altitude plants and animals are found in the Himalayas. In addition, this region is the centre of the origin for many crop and fruit tree species; as such, it is an important source of genes for its wild relatives. As a consequence of global warming, the present distribution of species in high-elevation ecosystems is projected to shift to higher elevations, although the rates of vegetation change are expected to be slow and the success in colonisation would be constrained by increased erosion and overland flows in the highly dissected and steep terrains of the Himalayan mountain range. The timberline varies between 3,300-3,600 m in the Western and 3,600-4,000 m in the Eastern Himalayas and represents an eco-tone between the sub-alpine and alpine zones. As a consequence of climate change, the timberline will be shifted, effecting the habitat and species of the area. The Himalayan region is home to several species of flora and fauna, including threatened medicinal plants such as *Taxus Wallichiana*, *Podophyllum hexandrum*, *Skimmia laureola*, *Valeriana Wallichii* and *Polygonatum verticillatum* besides the Himalayan musk deer, Himalayan tahr and Himalayan black bear. Much of these habitats and species would become extinct in the event of climate change.

---

14. Ravindranath *et al.* (2006). "Impacts of climate change on forest in India", *Current Science* 90 (3): 354-359, 10 February.
15. GoI, 2008 *op. cit.*

Impact on Health

Under climate change scenarios, as reported in IPCC AR4, there is likely to be a net increase in the geographic range of potential transmission of malaria and dengue—two vector-borne infections.[69] Malaria and dengue have already increased their geographical range to metropolises such as Mumbai.[70] Within their present ranges, these and many other infectious diseases would tend to increase in incidence and seasonality—although regional decline would occur in some infectious diseases. In all cases however, actual disease occurrence is strongly influenced by local environmental conditions, socio-economic circumstances and public health infrastructure. Changes in climate could alter the distribution of the important vector species and may also increase spread of such diseases to new areas. The studies conducted under NATCOM and GoI-DEFRA Assessments indicate that:[16,17,18]

- Malaria is likely to persist in Orissa, West Bengal and the southern parts of Assam, bordering north of West Bengal. It may shift from central India to southwestern coastal states of Maharashtra, Karnataka and Kerala. All northern states, including Himachal Pradesh and northeastern states of Nagaland, Manipur and Mizoram may become malaria-prone.
- The duration of the transmission windows is likely to widen in the northern and western states and shorten in southern states. The transmission window in Jammu and Kashmir and in Rajasthan may increase by 3-5 months. However, in Orissa and some southern states, a further increase in temperature is likely to shorten the transmission window by 2-3 months.
- If there is an increase of 3.8°C in temperature and a 7 per cent increase in relative humidity, the transmission window, i.e., months during which mosquitoes are active, will be open for all 12 months in 9 states in India.

Projected climate change will also be accompanied by an increase in heat waves, often exacerbated by increased humidity and urban air

16. Bhattacharya, Sharma *et al.* (2006). "Special Section: Climate change and India", *Current Science* 90(3): 314-95, 10 February.
17. GoI, 2008 *op. cit.*, p.16.
18. GoI, 2004 *op. cit.*, p.120.

pollution, which would cause an increase in heat-related deaths and illness episodes [7,17]. Exposure to higher temperatures and air pollutants appear to be significant risk factors for cerebral infarction and cerebral ischemia during the summer months. The evidence indicates that the impact would be the greatest among urban populations, affecting particularly the elderly and sick. Limited evidence indicates that increased summer deaths due to heat strokes would occur in some northern Indian states. Any increase in flooding will increase the risk of drowning, diarrhoeal and respiratory diseases, and of hunger and malnutrition. Influenza pandemics are an evolutionary certainty. But H5N1 notwithstanding, the next influenza pandemic may begin next month, next year or ten years from now. With more intense cyclones, devastating impacts would occur, particularly in densely settled populations with inadequate resources. A reduction in crop yields and food production because of climate change in some states will predispose food-insecure populations to malnutrition, leading to impaired child development and decreased adult activity. Socio-economic disruptions could occur in some regions, impairing both livelihoods and health. Overall, the adverse health impacts of climate change will be greatest in vulnerable lower income populations of India.

A range of social, institutional, technological and behavioural adaptation options need to be practiced to lessen the impact of climate change on human health. Adaptations could, for example, encompass strengthening of the public health infrastructure, health-oriented management of the environment (including air and water quality, food safety, urban and housing design and surface water management), and the provision of appropriate medical care facilities.

Based on a review of published work till date, the climate change induced impacts on human health in India by the year 2050 would be:

- Climate change will be accompanied by an increase in heat waves, often exacerbated by increased humidity, leading to an increase in heat-related deaths and illness episodes particularly among the elderly and sick in urban populations.
- Potential climate change impacts necessitate improved contingency planning and national guidelines to avoid disruptions in the health services in the event of rapid mutation/new strains of the deadly viruses.

- With more intense cyclones, devastating impacts of storm surges and floods would occur, particularly in densely settled populations, characterised by a policy cycle in which there is a flurry of activity when the event hits. Long-term planning is required to reduce human vulnerability to these inevitable events.
- Further strengthening of the public health and epidemiological surveillance would be required for health-oriented management of the environment (including air and water quality, food safety, urban and housing design and surface water management) and urban infrastructures (storm drains, water supply and waste management systems).

Impacts on Marine Biodiversity

India has 13 major wetland types occupying an area of 715, 819 sq km, including estuaries, lagoons, marshes, coral reefs, salt pans and ponds. Adverse impacts of climate change on Indian natural ecosystems such as coral reefs, mangroves and wetlands are also likely [68]. The biodiversity of coral reefs includes sea grasses, corals, several invertebrate groups, fishes, amphibians, birds nesting on reefs and mammals. The reefs of the Andaman and Nicobar Islands have the highest recorded diversity with 203 coral species, 120 algal species and 70 sponges in addition to fish, sea turtles, dugongs and dolphins. Violent storms in recent years have washed away nesting sites of Olive Ridley and Leatherback turtles on India's eastern and southern coastline.

India's 7,500 km coastline will be particularly hard-hit by storm surges and rise in sea levels, displacing millions, flooding low-lying areas and damaging economic assets and infrastructure. Frequent inundation of low-lying areas, drowning of coastal marshes and wetlands, enhanced erosion of beaches, more flooding and increased salinity of rivers, bays and aquifers in the coastal regions of India have occurred in recent years.

Projected climate variations and resultant rise in sea level would flood existing tidal habitats of an ecologically important, sensitive and productive nature. Climate-related stresses in coastal areas include loss and salinisation of agricultural land due to change in sea level, the likely changes in intensity of tropical cyclones and possibility of reduced productivity in coastal and oceanic fisheries. For a country with such a long and mythic self-identification with our plant and wildlife, the loss of our natural heritage will carry both socio-cultural as well as significant livelihood implications.

In India, the climate change induced impacts on coastal zones and marine ecosystem by 2050 would be:

- Climate change and rising sea levels tend to worsen the currently eroding coasts, turn the stable coasts into recession coasts. Sea level rise will accelerate coastal erosion by different coastal processes along the vast Indian coastline. Here, the relative sea level rise may be somewhat higher than the global mean by the mid-21st century. Coastal erosion will be enhanced as rising sea level allows higher wave and storm surge to hit the shore.
- The threats to the resilience of both human and environmental coastal systems are likely to be exacerbated by climate change. The projected sea level rise could inundate low lying areas, drown coastal marshes and wetlands, erode beaches, exacerbate flooding and increase the salinity of rivers, bays and aquifers.
- The rise in sea levels also threatens the ecologically sensitive areas such as dunes, beaches and coastal wetlands, including salt marsh habitats and mangroves along the Indian coastline. Projected increase in frequency and intensity of extreme weather will exert adverse pressure on our aquatic ecosystems, and existing habitats will be redistributed, affecting estuarine flora distribution.

If conservation measures receive increasing attention, large areas of the reefs along the Indian coastline could recover from the direct and indirect damage. However, if abnormally high sea surface temperatures continue to cause major bleaching and reduce the capacity of reefs to calcify due to increasing CO_2, most of human conservation efforts will prove futile.

Impact on Extreme Events

The future climate change projections for India indicate that some weather extremities will become more frequent, more widespread and/or more intense during the 21st century. More intense rainfall spells are also projected in a warmer atmosphere, increasing the probability of extreme rainfall and hence, flash floods in parts of India. A summary of the expected effects over the Indian subcontinent is listed in Table 2.2. Detailed analysis of the impact of climate change on extreme events, *viz.*, cyclones, floods, drought, etc., is covered in Chapter 3.

Table 2.2

Summary of Expected Effects on Extreme Events Over Indian Subcontinent

Phenomenon and Direction of Trend	*Likelihood in the 21st Century*
Over most land areas, warmer and fewer cold days and nights, warmer and more frequent hot days and nights.	Virtually certain (greater than 99% chance)
Warm spells/heat waves. Frequency increases over most areas.	Very likely (90–99% chance)
Heavy precipitation events. Frequency increases over most areas.	Very likely (90–99% chance)
Area affected by drought increases.	Likely (66–90% chance)
Intense tropical cyclone activity increases.	Likely (66–90% chance)
Increased incidence of extreme high sea level (excluding tsunamis).	Likely (66–90% chance)

Source: IPCC Interim Working Group I Report, April 2007; IPCC Synthesis Report, November 2007.

Conclusions

Many uncertainties remain—including anthropogenic aerosol burden—which need more studies; indeed, we expect that uncertainties will always remain. The nature and intensity of future greenhouse gas emissions depends on future decisions of governments and individuals, the speed of deployment of alternative energy systems, population sizes and lifestyles and many other factors. The global and regional climate models that simulate the role of these greenhouse gases in our atmosphere are yet imperfect. There are significant limitations in the ability of climate models to incorporate and reproduce important aspects of the Indian summer monsoon.

At present, we must admit that our institutional capability and experienced manpower to provide confident future projections of regional climate change scenarios and its intra-seasonal/inter-annual variability in space is limited, and more concerted and multi-institutional R&D is required to attain excellence in climate modelling efforts. Nonetheless, from what we know today, it could be concluded that:

- Over the Indian subcontinent, the annual mean area-averaged surface warming is projected to be between 2.0° and 3.0°C, and 3.5° and 5.5°C by the middle and end of the 21st century respectively. The warming would be more pronounced in winter than during the monsoon season.
- India may experience a decline in winter rainfall between 5 and 25 per cent, while only about a 10 per cent increase in area-averaged monsoon rainfall is projected by the end of the 21st century. The decline in winter rainfall over India is likely to be significant and may lead to droughts during the dry summer months.
- The date of the onset of the summer monsoon over central India would become more variable. More spells of intense rainfall are projected in a warmer atmosphere, increasing the probability of extreme rainfall and hence flash floods in parts of India.
- Shortage of drinking and irrigation water in drought-affected areas of India would cause food shortages, and in turn, human malnutrition. Decreased yields in non-irrigated wheat and in rice are very likely.
- The retreat of most glaciers is projected to accelerate and many small glaciers may disappear. For mountainous states, the risks of increased erosion, mudslides and other natural hazards can be anticipated with climate change.
- Life supporting food production in coastal areas of India is also changing and threatened, including the location of fish populations, and food security is at risk.

Recommendations

Our institutional capability and experienced manpower to provide confident future projections of regional climate change scenarios and its intra-seasonal/inter-annual variability in space is limited and more concerted and multi-institutional R&D is required to attain excellence in climate modelling efforts.

Some weather conditions and extremities will become more frequent, more widespread and/or more intense over the Indian subcontinent during the 21st century. More intense rainfall spells are also projected in a warmer atmosphere, increasing the probability of extreme rainfall events and hence, flash floods in parts of India. This calls for an appropriate disaster preparedness plan to effectively tackle such situations in densely populated urban centres.

Glacial runoffs in summer and river flows will also go down, leading to severe water shortages in north India in the long run. Sustainable consumption of freshwater needs to be practiced, to conserve available water resources, for maintaining biodiversity and ecological functions and services as strategies for responding to impacts of climate change on water resources.

The impacts of climate change on agricultural productivity and prices are estimated to result in trade deficit and income losses in India. A substantial decrease in cereal export trade in India would also require expansion of our infrastructures for efficient management.

Strengthening of the public health and epidemiological surveillance would be required for health-oriented management of the environment (including air and water quality, food safety, urban and housing design and surface water management) and urban infrastructures (storm drains, water supply and waste management systems).

The rise in sea levels could inundate low lying areas, drown coastal marshes and wetlands, erode beaches, exacerbate flooding and increase the salinity of rivers, bays and aquifers along the vast Indian coastline. All future developmental activities in the coastal regions of India should account for measures to counter such eventualities.

End Notes

1. Kripalani, R.H., A. Kulkarni, S.S. Sabade and M.L. Khandekar (2003). "Indian monsoon variability in a global warming scenario", *Natural Hazards* 29(2): 189-206.
2. Lal, Murari (2003). "Global climate change: India's monsoon and its variability", *Journal of Environmental Studies and Policy* 6 (1): 1-34.
3. Lal, Murari (2005). "Climate change—implications for India's water resources", in M.M.Q. Mirza and Q.K. Ahmad (eds.), *Climate Change and Water Resources in South Asia.* Leiden, The Netherlands: A.A. Balkema Publishers. pp.155-96.
4. Meehl, G.A. and T.F. Stocker (2007). "Global climate projections", in S. Solomon *et al.* (eds.), *Climate Change 2001: The Scientific Basis: Contribution of Working Group I to the Third Assessment Report of the Intergovernmental Panel on Climate Change.* Cambridge University Press.
5. Lal, M., B. Bhaskaran and S. K. Singh (1998a). "Indian summer monsoon variability as simulated by regional climate model nested in a global climate model", *Chinese* 22(1): 93-102.

6. Lal, M., G.A. Meehl and Julie M. Arblaster (2000). "Simulation of Indian summer monsoon rainfall and its intraseasonal variability", *Regional Environmental Change* 1(3 & 4): 163-79.

7. Lal, M., H. Harasawa and D. Murdiyarso (2001a). "Asia", in J.J. McCarthy *et al.* (eds.), *Climate Change 2001–Impacts, Adaptation and Vulnerability, Contribution of Working Group II to the Third Assessment Report of the Intergovernmental Panel on Climate Change.* Cambridge University Press. pp.533-90.

8. Lal, M., T. Nozawa, S. Emori, H. Harasawa, K. Takahashi, M. Kimoto, A. Abe-Ouchi, T. Nakajima, T. Takemura and A. Numaguti (2001b). "Future climate change: Implications for Indian summer monsoon and its variability", *Current Science* 81(9): 1196-1207.

9. May, W. (2004b). "Potential future changes in the Indian summer monsoon due to greenhouse warming: analysis of mechanisms in a global time-slice experiment", *Climate Dynamics* 22: 389-414.

10. Lal, M. (2002). "Global climate change: India's monsoon and its variability", Final Report under Country Studies Vulnerability and Adaptation Work Assignment with Stratus Consulting's Contract of the U.S. Environmental Protection Agency, pp.58.

11. May, W. (2004a). "Simulation of the variability and extremes of daily rainfall during the Indian summer monsoon for present and future times in a global time-slice experiment", *Climate Dynamics* 22(2-3): 183-204.

12. Mirza, M.M.Q. and Q.K. Ahmad (eds.) (2005). *Climate Change and Water Resources in South Asia.* Leiden, The Netherlands: A.A. Balkema Publishers.

13. Mirza, M.M.Q., R.A. Warrick, N.J. Erickson and G. J. Kenny (2005). "Are floods getting worse in the GBM Basins?" in Mirza and Ahmad Ibid. pp.55-76.

14. Giorgi, F. (2006). "Climate change hot-spots", *Geophys. Res. Lett.*, 33, L08707, doi:10.1029/2006GL025734.

15. Giorgi, F. and X. Bi (2005). "Regional changes in surface climate interannual variability for the 21st century from ensembles of global model simulations", *Geophys. Res. Lett.,* 32, L13701, doi:10.1029/2005GL023002.

16. Hasnain, S.I. (2002). "Himalayan galciers meltdown: Impacts on South Asian Rivers", *IAHS Pub.* No.274, pp.1-7.

17. IPCC (2007). *Climate Change Synthesis Report: A Contribution of Working Groups I, II and III to the Fourth Assessment Report of the Intergovernmental Panel on Climate Change.* Cambridge, UK and New York: Cambridge University Press. p.398.

18. IPCC (2007). *Climate Change 2007: The Physical Science Basis–A Contribution of Working Group I to the Fourth Assessment Report of the Intergovernmental Panel on Climate Change*, Cambridge, UK and New York: Cambridge University Press.

19. Emanuel, K. (2005). "Increasing destructiveness of tropical cyclones over the past 30 years", *Nature* (436): 686-88.

20. Christensen, J.H. and B. Hewitson (2007). "Regional climate projections", in S. Solomon *et al.* (eds.), *Climate Change: The Scientific Basis, Contribution of Working Group I to the Third Assessment Report of the Intergovernmental Panel on Climate Change.* Cambridge University Press.

21. Lal, M. (2001). "Tropical cyclones in a warmer world", *Current Science* 80(9): 1103-1104.

22. Knutson, T.R. and R.E. Tuleya (2004). "Impacts of CO_2 induced warming on simulated hurricane intensities and precipitation: Sensitivity to the choice of climate model and convective parameterization", *Jr. Climate* 17: 3477-95.

23. Walsh, K.J.E. (2004). "Tropical cyclones and climate change: Unresolved issues", *Climate Research* 27: 77-84.

24. Yohe, G. and K. Strzepek (2005). "Climate change and water resources assessment in South Asia: Addressing uncertainties", in Mirza, M.M.Q. and Q.K. Ahmad *op. cit.,* pp.77-102.

25. Trenberth, K. (2005). "Uncertainty in hurricanes and global warming", *Science* 308(5729): 1753-54.

26. Sridharan, S. and A. Muthuchami (2002). "Some salient features of the Bay of Bengal super cyclonic storm of October", *Mausam* 53(2): 237-41.

27. Lal, M., K.K. Singh, L.S. Rathore, G. Srinivasan, S.A. Saseendran (1998b). "Vulnerability of rice and wheat yields in NW-India to future changes in climate", *Agri. & Forest Meteorol.* 89: 101-14.

28. Aggarwal, P.K., P.K. Joshi, J.S.I. Ingram and R.K. Gupta (2004). "Adapting food systems of the Indo-Gangetic plains to global environmental change: Key information needs to improve policy formulation", *Environ. Sci. Policy* 373: 1-12.

29. Dawe, D., A. Dobermann, P. Moya, S. Abdulrachman, Bijay Singh, P. Lal, S.Y. Li, B. Lin, G. Panaullah, O. Sariam, Y. Singh, A. Swarup, P.S. Tan, and Q.X. Zhen (2000). "How widespread are yield declines in long-term rice experiments in Asia?" *Field Crops Res.* 66: 175-93.

30. Easterling, W. and P. Aggarwal (2007). "Food, fibre and forest products", in M. Parry *et al.* (eds.), *Climate Change, 2007: Impacts, Adaptation and Vulnerability, Contribution of Working Group II to the Fourth Assessment Report of the Intergovernmental Panel on Climate Change.* Cambridge University Press.

31. Bruinsma (ed.) (2003). *World Agriculture: Towards 2015/2030: An FAO Perspective.* Rome : Food and Agriculture Organization, United Nations and London: Earthscan.

32. Food and Agriculture Organization (2005). "Background document on special event on impacts of climate change, pests and diseases on food security and poverty reduction" 31st Session of the *Committee on World Food Security.* United Nations. May 23-25.

33. Faisal, I. M. and S. Parveen (2004). "Food security in the face of climate change, population growth, and resource constraints: Implications for Bangladesh", *Environmental Management* 34(4): 487-98.

34. Farooq, A.B. and A.H. Khan (2004). "Climate change perspective in Pakistan", Proceedings of the *Capacity Building APN Workshop on Global Change Research.* Islamabad, Pakistan. June 8-10. pp.39-46.

35. Hobbs, P.R. (2001). "Tillage and crop establishment in South Asian rice-wheat systems: Present and future options", *Journal of Crop Production* 4: 1-23.

36. Rosenzweig, C., A. Iglesias, X.B. Yang, Paul R. Epstein, and Eric Chivian (2001). "Climate change and extreme weather events: Implications for food production, plant diseases and pests", *Global Change & Human Health* 2(2): 90-104. Kluwer Academic Publishers.

37. Bhadra, B. (2002). "Regional cooperation for sustainable development of Hindu Kush Himalaya region: Opportunities and challenges", Keynote Paper presented at the *Alpine Experience: An Approach for other Mountain Regions*? Berchtesgaden, Germany. June 26-29.

38. Ali, M. and D. Byerlee (2000). "Productivity growth and resource degradation in Pakistan's Punjab: A decomposition analysis", *Economic Policy Working Paper* No. 2480. Washington D.C.: The World Bank.

39. Kaul, M.K. (ed.) (1999). *Inventory of the Himalayan Glaciers: A Contribution to the International Hydrological Programme.* Vedams Book Publishers.

40. Kataki, Palit K. (Ed.) (2001). *The Rice-Wheat Cropping System of South Asia: Trends, Constraints, Productivity and Policy.* Haworth Press.

41. Molden, D. and C. de Fraiture (2004). *Investing in Water for Food, Ecosystems and Livelihoods.* Blue Paper, *Discussion Draft*. Stockholm, Sweden.

42. Naithani, A.K., H.V. Nainwal, K.K. Sati and C. Prasad (2001). "Geomorphological evidences of retreat of the Gangotri glacier and its characteristics", *Current Science* 80(1): 87-94.

43. New Scientist (1999). "Flooded out: Retreating glaciers spell disaster for valley communities", *New Scientist: 18,* 5 June.

44. Pakistan Council of Research in Water Resources (2003). *Determination of Crop Water Requirement of Major Crops under Shallow Water Table Conditions.* Research Report-2.

45. Postel, Sandra L. (1998). "Water for food production: Will there be enough in 2025?", *BioScience* 48(8): 629-37.

46. Qureshi, A.S. (2005). "Climate change and water resources management in Pakistan", in M.M.Q. Mirza and Q.K. Ahmad (eds.), *Climate Change and Water Resources in South Asia.* Leiden, The Netherlands: A.A. Balkema Publishers. pp.197-230.

47. Gupta, S.K. and R.D. Deshpande (2004). "Water for India in 2050: First-order assessment of available options", *Current Science* 86(9): 1216-24.

48. Patz, J.A., D. Cambell-Lendrum, T. Holloway and J.A. Foley (2005). "Impact of regional climate on human health", *Nature* 438: 310-17.

49. Royal Society (2005). *Food Crops in a Changing Climate.* Report of a Royal Society Discussion Meeting held in April 2005, Policy Document 10(5). Available at *www.royalsoc.ac.uk*

50. Food and Agriculture Organization (2005). *Eradicating World Hunger: Key to Achieving the Millennium Development Goals.* The State of Food Insecurity in the World (SOFI). Rome.

51. Sharma, S. (2001). "The Ganges crisis: Government lacks strategic plan", in *Proceedings of the International Conference on Freshwater.* Bonn, Germany, 3-7 December.

52. Singh, P. and L. Bengtsson (2004). "Hydrological sensitivity of a large Himalayan basin to climate change", *Hydrological Process* 18: 2363–85.

53. Shrestha, A.B. (2004). "Climate change in Nepal and its impact on Himalayan glaciers", Paper presented at European Climate Forum Symposium on *Key Vulnerable Regions and Climate Change: Identifying Thresholds for Impacts and Adaptation in Relation to Article 2 of the UNFCCC.* Beijing, 27-30 October.

54. Krishnamurthy, V. and B.N. Goswami (2000). "Indian monsoon-ENSO relationship on interdecadal timescale", *J. Clim* 13: 579–95.

55. Shaman, J. and E. Tziperman (2005). "The effect of ENSO on Tibetan Plateau snow depth: A stationary wave teleconnection mechanism and implications for the South Asian monsoons", *Jr. Climate* 18: 2067-79.

56. Pai, D.S., V. Thapliyal and P.D. Kokate (2004). "Decadal variation in the heat and cold waves over India during 1971-2000", *Mausam* 55(2): 281-92.

57. Kalsi, S.R. and Pareek, R.S. (2001). "Hottest April of the 20th century over northwest and central India", *Current Science* 80(7): 867-73.

58. India Meteorological Department (various years). *Annual Reports, 2000, 2001, 2002, 2003.*

59. India Meteorological Department (various years). *Southwest Monsoons of 2002, 2003, 2003, 2004, 2005 and 2006: End of Season Reports*. Pune: India Meteorological Department, Government of India.

60. De, U.S., R.K. Dube and G.S. Prakasa Rao (2005). "Extreme weather events over India in the last 100 years", *Journal of the Indian Geophysical Union* 9(3): 173-87.

61. Kharin, V.V. and F.W. Zwiers (2000). "Changes in extremes in an ensemble of transient climate simulations with a coupled atmosphere-ocean GCM", *J. Climate* 13: 3760–80.

62. Daniel A. Cenderelli Ellen and E. Wohl (2003). "Flow hydraulics and geomorphic effects of glacial-lake outburst floods in the Mount Everest region, Nepal", *Earth Surface Processes and Landforms* 28(4): 385-407.

63. WWF (2005). *An Overview of Glaciers, Glacier Retreat, and Subsequent Impacts in Nepal, India and China.* Technical Report. Nepal.

64. Dyurgerov, Mark B. and Mark F. Meier (2005). "Glaciers and the changing Earth system: A 2004 Snapshot", *Occasional Paper* No. 58, Institute of Arctic and Alpine Research.

65. Slingo, J.M., A.J. Challinor, B.J. Hoskins and T.R. Wheeler (2005). "Introduction: Food crops in a changing climate", *Phil. Trans. Royal Soc.* B 360: 1-7.

66. Swaminathan, M.S. (2003). "Count down from 2007", *The Hindu Magazine*, 27 April. p.A1+.

67. Sodhi, N.S., L.P. Koh, B.W. Brook and P.K.L. Ng (2004). "Southeast Asian biodiversity: an impending disaster", *Trends in Ecological Evolution* 19: 654-660.

68. Sukumar, R., K.G. Saxena and A. Untawale (2003). "Climate change impacts on natural ecosystem", in P.R. Shukla, S.K. Sharma, N.H. Ravindranath, A. Garg and S. Bhattacharya (eds.), *Climate Change and India: Vulnerability Assessment and Adaptation*. Hyderabad : Orient Longman. pp. 266-290.

69. Mitra, A, S. Bhattacharya, R.C. Dhiman, K.K. Kumar and C. Sharma (2004). "Impact of climate change on health: A case study of malaria in India", in P.R. Shukla, S.K. Sharma, N.H. Ravindranath, A. Garg and S. Bhattacharya (eds.), *Climate Change and India: Vulnerability Assessment and Adaptation*. Hyderabad: Orient Longman. pp. 360-88.

70. Singh, N., A.C. Nagpal, A. Saxena and A.P. Singh (2004). "Changing scenario of malaria in central India, the replacement of Plasmodium vivax by Plasmodium falciparum (1986-2000)", *Trop. Med. Int. Health* 9: 364-71.

71. Unnikrishanan A.S., Shankar D. (2007). "Area sea levels trends along the North Indian Ocean coasts consistent with global estimates?", *Global and Planetary Change* 57 301.

72. GoI (2008). *National Action Plan on Climate Change (NAPCC).* Prime Minister's Council on Climate Change. Available at *www.pmindia.nic.in*

73. GoI (2004). *India's Initial National Communication (NATCOM) to the United Nations Framework Convention on Climate Change (UNFCCC).* Ministry of Environment and Forests (MoEF).

74. GoI/DEFRA (2005a). "Indian Climate Change Scenarios for Impacts Assessments, Key Sheet 2—Indian Institute of Tropical Meteorology (IITM), Pune, India", in *Investigating the Impacts of Climate Change in India*. Report by Department of Environment, Food and Rural Affairs (DEFRA), UK and Ministry of Environment and Forests (MoEF). Available at *http://www.defra.gov.uk/environment/climatechange/internat /devcountry/ india2.htm*

75. GoI/DEFRA (2005b). "Climate Change Impacts on Sea Level in India. Key Sheet 4—National Institute of Oceanography, Goa, India", in *Investigating the Impacts of Climate Change in India*. Report by Department of Environment, Food and Rural Affairs (DEFRA), UK and Ministry of Environment and Forests (MoEF). Available at *http://www.defra.gov. uk/environment/climatechange/internat/devcountry/india2.htm*

76. Special Section (2006). "Climate change and India", *Current Science* 90(3): 314-95, 10 February.

77. Ravindranath N.H., Joshi *et al.* (2006). "Impacts of climate change on forest in India', *Current Science* 90(3): 354-59, 10 February.

78. Bhattacharya S., C. Sharma *et al.* (2006). "Climate change and malaria in India". *Current Science* 90(3): 369-75, 10 February.

79. Kumar, Rupa *et al.* (2006). "High-resolution climate change scenarios for India for the 21st century", *Current Science* 90(3): 334-45, 10 February.

3 Climate Change and Disasters

Sunil Chauhan

Introduction

Climate change and disasters are fast emerging as the most defining challenges of the 21st century. Natural disasters are increasing in terms of frequency, complexity, scope and destructive capacity. Hitherto, disasters were linked to climate change through short-term natural variability, manifesting in extreme weather conditions such as cyclones, storms, floods, droughts, heat waves, windstorms and other natural hazards with potential for catastrophic loss of human lives, damages to infrastructure and environment. However, with climate change manifesting itself at an unprecedented rate with increased variability and frequency of extremity, these two processes have the potential to generate destructive forces which could cause colossal disasters unless urgent, radical and resolute steps towards mitigation are not implemented.

This chapter analyses the impact of climate change on the risk of natural hazards in India and proposes preparedness strategies and emergent policy imperatives.

Linking Climate Change and Disasters

The UN defines disaster as "a serious disruption of the functioning of the community or society, causing widespread human, material, or environmental losses which exceed the ability of the affected community/society to cope using its own resources."[1] Climate has always been linked with disaster, hitherto through climate variability manifesting in extreme weather conditions with potential for catastrophic loss of human lives, damage to infrastructure and environment. These short-term climate fluctuations and extreme weather conditions have been

1. UNDHA (1992).

the most frequently occurring hazards and in combination with social vulnerability, have been responsible for a majority of disasters worldwide. Such natural disasters resulting from climate variability and other climatic and meteorological causes are commonly referred as "climate disasters" or "hydrometeorological disasters" in disaster studies.[2]

Climate Disasters are Chronic and Increasing

Climate disasters have been a recurring theme in human history and are on a rising trend.

- Trends—20th Century: When compared to geophysical disasters, climate-related disasters were seven times as frequent, accounted for nine times as many deaths and caused economic losses that were three times higher.[3]
- The Last Three Decades and Early 21st Century: Over the past three decades, climate-related natural disasters occurred five times as frequently, killed or affected 70 times as many people and caused twice as much damage worldwide as did earthquakes and volcanoes. In the past decade, weather-related natural hazards have been the cause of 90 per cent of natural disasters and 60 per cent of related deaths and have been responsible for 98 per cent of the impacts on disaster-affected populations, the majority in developing countries.[4] Between 2000 and 2004, an average of 326 climatic disasters was reported each year. Some 262 million people were affected annually for 2000-2004, more than double the level in the first half of the 1980s.[5]
- Climate Disasters Statistical Review 1987-2006: The global review of the statistics of climate disasters by CRED indicates that climate

2. The Centre for Research on the Epidemiology of Diseases (CRED), Brussels categorises the disasters resulting from climatic variability and other climatic and meteorological causes as 'hydrometeorological disasters' (floods, landslides, mudflows, avalanches, tidal waves, windstorms, including typhoons, cyclones, hurricanes, storms, winter storms, tropical storms and tornadoes, droughts, extreme temperatures and complex disasters associated with drought) as distinct from 'geological disasters' (earthquakes, volcanic eruptions and tsunamis). See CRED (2007).
3. UNDP (2007b).
4. IFRC (2005).
5. UNDP (2007a).

disasters[6] are increasing in terms of frequency, complexity, scope and destructive capacity. A comparison of the occurrences of climate disasters from 1987-2006 with averages for 1987-1998 and 2000-2006[7] is at Figure 3.1.

Figure 3.1

Occurrence of Hydrometeorological Disasters: 1987 with Averages for Periods 1987-1998 and 2000-2006

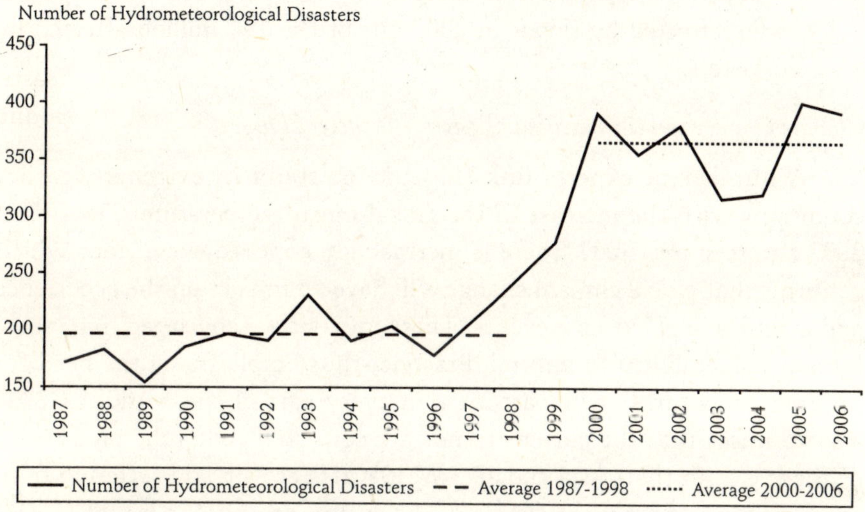

Source: CRED, 2007.

The statistics from Figure 3.1 indicate that during the period 1987-2006, climate disasters showed a significant increase. For 1987-1998, the average number of climate disasters reported was 195; for the years 2000-2006, this number increased by 187 per cent to an average of 365. The years 1998-99 and 2000 reflect a turning point and an escalation in the number of disasters reported, with 1999 showing the most dramatic inflection. Accordingly, for purposes of the comparing the two periods, data for 1999 was excluded.

- Disaster Figures for 2007: The most recent data from CRED for 2007 reveals a marked increase in the number of climate disasters,

6. In this chapter, the statistics for 'hydrometeorological disasters' (categorised by CRED as resulting for climatic variability and other climate causes) is used to represent 'climate disasters'.

7. CRED (2007).

especially floods, compared with the average of the last seven years, with Asia as the hardest hit continent.[8] Eight out of the ten countries with the highest disaster deaths of 2007 were in Asia, with 4,234 killed in Bangladesh by cyclone Sidr in November 2007 and more than 3,000 fatalities from severe floods in Bangladesh, India, North Korea and China. The number of people affected by disasters continued to increase and floods remain the main disaster that affects populations in the world. More than 164 million people were affected by floods in 2007 out of the 197 million affected by disasters.

Climate Change and Increasing Disasters—Is there a Link?

A number of experts link the current trend in extreme weather conditions with the increase in the global mean temperature. The CRED 2007 report states that, "there is increasingly conclusive evidence which confirms that global climate change will have an impact on the occurrence and magnitude of extreme events. These impacts are envisaged to increase human vulnerability to natural disasters, thus emphasizing the need for improved measures of preparedness in every part of the world".[9] CRED also surmises that the current trends are consistent with the predictions of the IPCC, in that, Asia and also West Africa are already suffering from more severe and frequent floods.[10] The future projections by IPCC AR4 also indicate that there is "increased confidence that some weather events and extremes will become more frequent, more widespread and/or more intense during the 21st century and impacts due to altered frequencies and intensities of extreme weather, climate and sea level events are very likely to change".[11]

Notwithstanding the predictions of IPCC 2007 AR4, supported by CRED through trend analysis of rising climate disasters, a direct attribution to climate change is not possible, as every weather condition is the product of random forces and systemic factors. However, climate change will steadily increase the exposure of the poor and vulnerable to climate shocks

8. UN/ISDR (2008/01)
9. Ibid.: 14.
10. Guha (2008).
11. IPCC (2007).

and place increased pressure on coping strategies and adaptive capacities of societies by resource and environment degradation. Current evidence points very clearly that climate change is creating systemic conditions for extremities of weather and will increase the risk of exposure to climate disasters through increasing societal vulnerabilities. Shifts from long-term averages could raise the level of background stress, thereby increasing vulnerability to acute environmental stresses. The character and frequency of the acute stresses themselves could change, manifesting as extreme events. Lastly, climate change could exacerbate the scope and intensity of chronic hazards and contributing elements associated with such stresses by increasing vulnerability because of increasing background environment stress.[12]

Disaster Vulnerability Profile of India

India is among the world's most disaster-prone countries. India supports 1/6th of the world's population on just 2 per cent of its landmass. Nearly 59 per cent of India's land area is prone to earthquakes of moderate to very high intensity, over 40 million hectares (12 per cent of land) is prone to floods, close to 5,700 kms of its 7,516 km coastline (about 80 per cent) is cyclone-prone and exposed to tsunamis and storm surges, 2 per cent of land is landslide-prone and 68 per cent of India's arable land is affected by droughts. Of the 35 states and union territories, as many as 27 are disaster-prone.[13] Most disasters in India are water-related.

Lack of availability and access to technological and financial resources coupled with high dependence on climate sensitive sectors—agriculture, fisheries, forestry—have made India highly vulnerable to climate change. A large and growing population, a densely populated and a low-lying coastline, and an economy closely tied to its natural resource base, further intensifies this vulnerability.

Present Disaster Trends in India and Future Projections

The global disaster trends are alarming, singularly so in India. The figures for 2006 in the CRED 2006[14] report indicate that Asia remains

12. Hultman and Alexander (2006).

13. GoI (2004a).

14. CRED (2007) *op. cit.:* 13.

the continent most hit by disasters, with over 44 per cent of all reported disasters. At aggregate country level, India ranks third with 21 significant disasters recorded (China recorded 38, followed by the US which recorded 31). In terms of victims, India ranked third with 7.3 million victims after China (88 million) and the Philippines (8.6 million). China, the US and India also remained the countries reporting the highest damages (China—US $ 13.5 billion, the US—US $5 billion and India—US $3.3 billion).[15] The available data for 2007 also has India reporting the third highest number of significant disasters, *viz.*, 18 as against 22 reported by the US and 20 by China. In terms of victims, India ranked the second highest recording 1,103 deaths, mainly due to floods, as against Bangladesh which recorded 5,534 (Cyclone Sidr and floods).[16] A scrutiny of the records of the past 100 years also indicates that India figures among the first 10 in the world in terms of fatalities and economic losses in a variety of disasters.

Likely Impact of Climate Change on Disasters in India

These trends are likely to exacerbate in future with climate change. Its consequences include a rise in sea level, threatening areas such as the densely populated Ganges delta, changes in the patterns of monsoon rains, the melting of the glaciers in the Hindu Kush-Karakoram-Himalaya region (crucial for the water supply in the dry seasons) and the foreseeable increase in heavy rain events and intensity of tropical cyclones.[17] The many glaciers located across the 2,400 km stretch of the Himalayan range are at the epicentre of an emerging crisis.

Climate change is also likely to threaten India's food security, increase water stress, result in rising sea levels and increase occurrences of diseases, especially malaria. Average temperatures are rising and dramatic and variable changes to extreme weather conditions are occurring due to climate change. Though it is not possible to predict the future frequency or timings of extreme events, there is consensus that the risk of drought, flooding and cyclone damage is increasing and will continue to do so.

The impacts will vary from region to region. The effect of possible changes in terms of intensity of the monsoons will be particularly sensitive

15. Ibid.: 33-34
16. ISDR/CRED (2007).
17. Parry, Canziani, Palutikof and Co-Authors (2007).

because large parts of India receive a majority of their annual precipitation during the summer monsoon rains, which already varies noticeably in different regions.[18] The summer monsoon is crucial to the total annual precipitation of the Indian subcontinent. The effect that global warming will have on the Indian monsoon is still unclear but increased variability in the monsoon rains is a probability, which would impact on floods and droughts.

Floods

In India, 40 million hectares are prone to floods with 8 million hectares being affected every year. The Brahmaputra and the Gangetic basins are the most flood-prone areas. The other flood-prone areas are the northwest region of west flowing rivers such as the Narmada and the Tapti, Central India and the Deccan region with major rivers flowing like the Mahanadi, the Krishna and the Cauvery.[19] Flooding is the largest cause of economic damage and loss of human lives in India. The population affected by floods in 2005 was 32.03 million and 1,504 lives were lost. In addition, about 96,713 livestock have been killed and 1,683 houses damaged. The loss to crops has been to the tune of Rs. 4,600.7 million (US $103 million), and the damage to public utilities is Rs. 3,772.48 million (US $5 million).[20] In 2006, India recorded the highest economic damages from floods in the world with losses amounting to US $3,390 and with over 40 lakh people affected (second to China, where over 41 lakh people were victims of floods).The floods in 2007 have also taken a significant toll with over 1,500 deaths, being the second highest after Bangladesh.[21]

Climate change will influence stream flow patterns through changes in the precipitation, changes in volume and timing of melting snow from glaciers, and changes in the type of precipitation, snow or rain, in the Himalayan region. Apart from changes in snowmelt and precipitation in the Himalayas, changes in the pattern, intensity and frequency of rainfall will have a significant impact on floods. Climatic changes could result in more frequent high-intensity rainfall. A preliminary assessment

---

18. Murari *et al.* (2001).
19. GoI (2002).
20. Kumar (2006).
21. ISDR/CRED (2007). *op. cit.*

undertaken by Indian scientists of the Indian Institute of Technology (IIT), Delhi, as part of the National Communication (NATCOM) on climate change project under the Ministry of Environment and Forests (MoEF), using the HadRM2 daily weather data to determine the spatial-temporal availability of the river systems in India, has indicated that the severity of floods under the projected climate change is likely to intensify.[22] The study based on a combination of changes in mean precipitation (due to climate change) and in the co-efficient of variation in runoff shows that Western India may experience very high river discharges more frequently than it does at present.

CYCLONES

India has a 7,516 km long coastline with two cyclone seasons, during the southwest and northeast monsoons. Cyclones have been observed to be more frequent in the Bay of Bengal than in the Arabian Sea. Consequently, the states of West Bengal, Orissa, Andhra Pradesh and Tamil Nadu along the Bay of Bengal are the most affected as compared to Maharashtra, Kerala, Karnataka and Gujarat which are along the Arabian Sea. Historically, the states that have experienced the greatest number of cyclones are Orissa and Andhra Pradesh and the Andaman and Nicobar Islands in the Bay of Bengal; the Lakshadweep Islands in the Arabian Sea are also threatened by cyclones. Most of the destructive cyclonic storms usually occur during the transition periods: pre- (April-May) and post- (September-December) monsoon, the latter being the most active period. The coastal districts of Orissa, Andhra Pradesh and Gujarat are most prone to the impacts of cyclone.[23] About 2.3 storms form on an average during a year.[24] Most casualties are caused by tidal waves, storm surges and torrential rains.

The East Coast of India, which lies in the path of tropical hurricanes from the Gulf of Bengal, is particularly at risk of being damaged by storms and floods.[25,26] The projections by the National Institute of Oceanography

22. Gosain and Rao *et al.* (2006).
23 GoI (2006).
24. GoI (2004b).
25. Emanuel (2005).
26. IPCC (2007).

(NIO) under the Council of Scientific and Industrial Research (CSIR), Government of India on the impacts of climate change on sea level (to assess the degree to which mean sea level and the occurrence of extreme events may change) showed an increased occurrence of cyclones in the Bay of Bengal, particularly in the post-monsoon period, along with increased maximum wind speeds associated with cyclones and a greater number of high surges under climate change.[27] In addition, the strength of tropical cyclones, which represent a threat to the eastern coast of India and to Bangladesh, could increase. The risk to these areas will be aggravated by the rising sea level.[28]

Droughts

Along with floods, India also suffers acute water shortage. Of the net area sown in the country, 68 per cent is prone to drought and of this, 33 per cent is chronically drought-prone, receiving rainfall of less than 750 mm per annum, while 35 per cent receives rainfall between 750-1,125 mm per annum.[29] The steady shrinking of the Himalayan glacier ranges will drastically cut down water availability in the downstream plains of Uttar Pradesh and Bihar. India's initial National Communication to the United Nations Framework Convention on Climate Change (UNFCC) projects that Luni, the west flowing river of Kutchh and Saurashtra, occupying about one-fourth of the area of Gujarat and 60 per cent of Rajasthan, are likely to experience acute physical water scarcity. The river basins of Mahi, Pennar, Sabarmati and Tapi are also likely to experience constant water scarcities and shortages.[30]

Landslides

Landslides affect large areas of the country every year during monsoons. They can also be caused by earthquakes. The areas that suffer from landslide hazards are located in the hilly tracts of the Himalayas, northeast India, Nilgiris, Eastern Ghats and Western Ghats. With the melting of the glaciers in the Hindu Kush-Karakoram-Himalaya region and

27. DEFRA/GoI (2005b).
28. German Advisory Council on Global Change (WBGU) (2006).
29. GoI (2002). HPC Report. *op. cit.*
30. GoI (2004b). NATCOM 2004. *op.cit.:* vii.

the foreseeable increase in heavy rain and intensity of tropical cyclones,[31] the incidences of landslides are likely to increase in vulnerable hilly areas.

The Coastal Zone

The coastal zone, a highly fragile ecosystem, is an important and critical region for India. It is densely populated with the Arabian Sea in the west and the Bay of Bengal in the east. The area occupied by the coastal districts is around 3,79,610 sq km, with an average population density of 455 persons per sq km, which is about five times the national average of 324 persons per sq km. Under the present climate conditions it has been observed that a rise in sea level (0.4-2.00 mm) along the Gulf of Kutchh and the coast of West Bengal is the highest. Along the Karnataka coast however, there is a relative decrease in sea level.[32]

Future climate change in the coastal zones is likely to be manifested through worsening of some of the existing coastal zonal problems like erosions, flooding, submergence and deterioration of coastal ecosystems, such as mangroves and salinisation. In many cases, these are caused by—or exacerbated by—rising sea levels and tropical cyclones. The key climate-related risks in the coastal zone include tropical cyclones, sea level rise and changes in the temperature and precipitation. A rise in sea level is likely to have significant implications on the coastal population and agricultural performance of India. A one metre sea level rise is projected to displace approximately 7.1 million people in India and about 5,764 sq km of land will be lost, along with 4,200 km of road.[33]

The diverse impacts expected as a result of sea level rise include land loss and population displacement, increased flooding of low lying coastal area and loss of yield and employment resulting from inundation and salinisation. Damage to coastal infrastructure, aquaculture and coastal tourism, due to erosion of the sandy beaches is also likely. The extent of vulnerability, however, depends not just on the physical exposure to sea level rise and the population affected but also on the extent of economic activity in the areas and capacity to cope with the impact.

---

31. Parry, Canziani, Palutikof and Co-Authors (2007).
32. GoI (2004b). NATCOM 2004. *op.cit.*: ix.
33. Ibid.: x.

A vulnerability assessment carried out of the Indian coastal areas based on their susceptibility to hazards (frequency of landing of severe storms on coastline), exposure (population density and housing index) and likely impacts of climate change (cumulative per lakh population and per event deaths) indicated that[34]:

- Jagatsinghpur and Kendrapara (earlier parts of Cuttack) in Orissa, Nellore in Andhra Pradesh, Nagapattinam in Tamil Nadu, Junagadh and Porbandar districts in Gujarat are highly vulnerable to impacts of climate change.
- North 24 Parganas, South 24 Parganas in West Bengal, Baleshwar and Bhadrak in Orissa, Srikakulam, East Godavari, Guntur, Krishna in Andhra Pradesh, Thiruvallur, Cuddalore, Ramanathapuram in Tamil Nadu are moderately vulnerable.
- All other coastal districts were classified as less vulnerable.

Natural ecosystems such as mangroves, grasslands and coral reefs are also likely to be affected by climate change. Rise in sea level would submerge mangroves as well as increase the salinity of wetlands. The projected sea level rise of 0.1-0.9 m seems within the ability of the Sunderbans mangrove ecosystem which presently faces tidal amplitudes up to 5 m. This may not be true for other mangroves such as the Pichavaram and Muthupet where tidal amplitudes are much lower at 0.64 m.[35] However, increased snow melt in the Western Himalayas could bring large quantities of freshwater into the Gangetic delta. This would have significant consequences on the composition of Sunderbans mangroves' species, favouring species that have least tolerance to salinity.

What is at Stake?

India is a leading global player of the future. India's economy is growing at a remarkable rate and if this trend is sustained, economic projections indicate that by 2035, India would become the world's third largest economy after China and the US.[36] Much would however, depend upon how she manages the emerging challenge of climate change and

34. Patwardhan *et al.* (2003).
35. Shukla *et al.* (2003).
36. DCDC (2007).

increasing disasters. For countries with developing economies like India, the financial setbacks which disasters inflict can be ruinous, as disasters disrupt short-term financial and economic management, simultaneously necessitating substantial realignment in spending plans, adjustments in economic targets and shifts in economic policy, in contrast to the developed counterparts, and have negative longer term consequences for economic growth, development and poverty reduction. More than 95 per cent of deaths caused by disasters occur in developing countries and the losses in developing countries are 20 times greater (as percentage of GDP) than in industrial countries because of limited coping mechanisms, prevention, mitigation and adaptation capabilities. Economic losses due to disasters in India were estimated to be Rs. 28,678 crore for the period 1994-1998, and rose to Rs. 47,464 crore in 1998-2003. In 2005 alone, disasters in India caused direct losses approaching Rs. 87,500 crore—more than the defence budget for the year which was Rs. 83,000 crore. Estimates for 2006 and 2007 are not yet available but are anticipated to be well over the Indian defence budget. So much of India's anticipated economic growth and emergence as a leading global power would depend on how India deals with disasters.

While India is a high growth economy, there is great inequity and a large human development backlog. Around 28 per cent of the population—some 320 million people—live below the poverty line.[37] Superimposing the incremental risk of climate change on this large development deficit would compromise the ambition of 'inclusive growth' set out in India's Eleventh Five Year Plan.

India is already battling high population and rapidly increasing urbanisation. The population of India is forecast to be over 1300 million by 2015. The urban population is growing faster than the rural population. The coastal mega cities of Chennai (2005: population 6.9 million), Kolkata (14.3 million) and Mumbai (18.2 million) lie generally only a few metres above sea level. The high population pressure combined with increasing cyclone intensity and sea level rise as a result of climate change will put millions of people at risk of being hit by storm and flood disasters.

With rapid development of coastal areas, industrialisation and urbanisation, more populations are becoming vulnerable to climate-

37. UNDP (2007a).

associated calamities and many have no choice but to move to safer places. It is estimated that approximately 142 million people may inhabit coastal India in 2050 and India's total number of flood zone migrants alone could be anywhere between 20 and 60 million, with 30 million as a conservative figure.[38]

Nearly 700 million rural people in India directly depend on climate-sensitive sectors (agriculture, forests and fisheries) and natural resources (water, biodiversity, mangroves, coastal zones and grasslands) for their subsistence and livelihood. Under changing climate, the food security of the country might come under threat. In addition, the adaptive capacity of dryland farmers, forest and coastal communities is low. Climate change is likely to impact all the natural ecosystems as well as health and socio-economic systems, as indicated by NATCOM.

Further, climate change, disasters and security are intrinsically linked. Major disasters pose many new threats and challenges to national security and stability. The sudden changes brought on by natural disasters exacerbate problems that people face daily and could challenge national security by heightening conditions for conflict such as inequality, resource scarcity, social grievances, political tension and distortions. Disasters could also destroy main political and social institutions, threatening political stability. The resulting increased competition for scarce resources, exacerbated inequality, changing power relations between individuals, groups and organisations could create conflict. The earthquake in Guatemala in 1976 caused a wave of destruction in the rural mid-section of the country that eventually brought an abrupt and violent change of leadership from the 21-year old military government. Similarly, Bangladesh was formed after angered victims of the 1970 typhoon instigated civil war in East Pakistan. Thus, a major disaster is a potential security threat.

Lastly, both climate change and disasters will play out in future against the backdrop of other global trends and developments. Both are largely anthropogenic processes, fuelled largely through development. Development processes are currently largely associated with risk accumulation and not risk reduction. During the next 30 years, every aspect of human life will change at an unprecedented rate, throwing

38. Myers (1993).

up new features, challenges and opportunities. The challenge of climate change and increasing disasters will have to be addressed in the backdrop of the key drivers of change, *viz.*, economic growth, rise in population, resource competition, changing demographics, increasing diseases, mass displacement and environmental impacts.

Recommendations

With climate change assuming ominous proportions, climatic disasters could assume gigantic proportions, which would threaten the security of the nation. Given the lack of resources and access to technology and finances, India has limited capacity to develop and adopt strategies to reduce their vulnerability to changes in climate. To manage these climate change-induced disasters, the country needs to have improved scientific understanding, capacity building, networking and broad consultation processes across every section of the society. Important policy recommendations are enumerated in the succeeding paragraphs.

Integrated Risk Management Framework

Climate change and disasters are fast emerging as the most defining challenges of the 21st century. The possibility of these two forces coalescing and generating greater destruction forces imposes imperatives for a comprehensive integrated risk management framework. In India, disaster management is being addressed by the National Disaster Management Authority (NDMA), which is the apex body for addressing policy issues and for laying guidelines. Climate change is being addressed at various levels, by the MoEF, the MoST, MEA as well as by the PMO. There is wide divergence among these entities. There is thus, an urgent need to bring climate change and the communities faced with such disasters onto a common platform, and develop an integrated risk management framework or model, to deal with the challenges of climate change and increasing disasters. An integrated approach would mean:

- Better coordination among climate change, disasters and development communities;
- Even-handed attention to the reduction of greenhouse gases and of the risks associated with climate change, including through enhanced disaster management; and

- Improved conceptual and methodological approaches to understand and respond to local manifestations of disasters while simultaneously addressing underlying complex and partly global processes.

Adaptation and Mitigation

Both adaptation and mitigation have been key levers in disaster risk reduction and climate change strategies. An integrated risk management framework needs to build on these two vital strategies. However, adaptation, in comparison to mitigation, has been the neglected stepchild in both disaster risk reduction and global negotiations and debates over climate change, mainly due to the complexity of separating adaptation from other socio-economic, environmental and development issues. Mitigation (structural, reduction of greenhouse gases, etc.) with clear visible manifestation of actions has been the easier chosen option. India has many mitigation projects for disasters, *viz.*, earthquake, flood and drought mitigation but the issue of adaptation is largely addressed only during discussions. This must change. The key levers of adaptation include poverty reduction, progressive change in economic structures, shifting away from primary (agriculture) livelihoods into (secondary and tertiary) knowledge-based forms of economic activity that are less vulnerable to direct impacts of risks, changes in land use and cropping patterns, etc., and enhancing the resilience of people. Adaptation will involve coping with climate change in the context of several factors that influence vulnerability. The Government of India has many policies in place that are directed at enhancing the adaptive capacity. However, implementation of these at the local level has been poor, as manifested in the slow eradication of poverty, inequity, etc., and the low human development index.

From Structural to Non-Structural Mitigation: The Ninth Five Year Plan saw the beginning of a gradual shift away from purely structural measures towards other forms of non-structural mitigation. In the Ninth Plan (1997-2001), it was observed that in addition to the progress made in implementing structural flood protection measures, flood forecasting and warning systems had played a great role in mitigating the loss of life and enabling the protection of moveable property. However, notwithstanding the growing understanding, the response of governments to floods and droughts (as evidenced from the Flood and Drought Response Synthesis) reveals that investment strategies focus on structures and other physical interventions designed to increase control over water availability and flow. These investments are intended primarily to increase the physical availability of water during dry periods. In the case of floods, most investments are directed towards the construction and maintenance of embankments and other flood control structures. Both floods and droughts, in the practical sense of investment, are seen as external events that can be controlled through the construction of physical structures. Concepts of flood and drought-proofing exist and do include the recognition of wide measures related to the vulnerability of livelihoods. Translating such concepts into governmental programmes that actually direct significant investment into something other than the construction of water control structures has not, however, occurred on a widespread basis. Lessons from past efforts towards flood control indicate that structural actions often increase long-term vulnerability. There has been a gradual shift from purely structural measures to non-structural measures for flood mitigation. This is reflected in the Tenth Five Year Plan of the Government of India, although much of the shift still remains on paper and awaits implementation.

Future Climate Change Disaster Scenarios—Impact Assessment

Practical strategies in an integrated framework of risk management would require reliable impact assessment of future risk scenarios. Impact assessment of the climate change-disaster nexus is complex as it includes geophysical and socio-economic aspects. However, tools used for these assessments need to be continuously evolved. Climate change research, monitoring and modelling is an emerging area in India and the assessments are in their nascent stage. The preliminary assessments are mostly sectoral in dimension, which do explicitly look at the inter-sectoral connection. Identifying and using appropriate sectoral tools and an integrated assessment approach with adequate data inputs, can lead to improved assessments with reduced uncertainties. There is a need to enhance technical and institutional capacity. Though India has a large science and technology institutional base in many areas relevant to climate change, these have to develop a shared vision, an integrated approach and networking for a synergistic assessment relevant to policy-making.

Climate Change/Disaster Risk Indices (CCDRI)

Quantitative indices are an important tool in risk management. International initiatives and research on developing risk indices for disaster and climate sensitivity exist, though in isolation. The UNDP has developed a Disaster Risk Index (DRI) with a global level of observation and a national level of resolution, allowing comparison between countries with respect to three types of hazards, *viz.*, earthquakes, cyclones and floods (India faces the highest risk, prone to floods with China a close second; for cyclones India ranks second with China leading, and for earthquakes India is ranked 10th).[39] With the confluence of climate change and disasters, development of climate change-induced disaster risk indices need to be initiated. The Tyndall Centre for Climate Change in the UK has commenced an initiative on this and is in the process of developing a methodology for climate change/disaster risk index at the global level. A similar initiative must commence from our National Steering Committee on Climate Change.

Climate Change Modelling

India has presently evaluated the outputs of ten General Circulation Models (GCMs) and one regional model, *viz.*, HadRM2, with a resolution of 50×50 km. However, regional climate modelling has to be strengthened and it is desirable to acquire/develop and set up more Regional Climate Models (RCM) to ascertain their simulation accuracy for the current Indian climate, and then use the appropriate RCM for more robust projections.

Vulnerability Atlas

A vulnerability map of India has been developed by the Building Material Council and Technology Promotion (BMCTP) showing multi-hazard zones based on earthquakes, wind and cyclones and floods. This aids in preparedness, quick risk assessment and action, is an invaluable tool for proactive planning, and has given enough warning regarding possible hazards that might get converted into disasters in the future. The vulnerability atlas needs to be revised in the context of the climate change projections and extreme weather conditions that India has faced in the last few years and also in the past on an intermittent basis.

39. UNDP (2004).

Development as a Tool for Reducing Risk

Climate change and disasters interface with diverse social and natural processes and consequently with the developmental process. The conventional view has been to consider disasters and climate change as a barrier to development and simultaneously, developments as a threat to climate change. However, development can be the driving force to overcome challenges and risks of climate change. The Indian national perspectives on sustainable development are consistent with this view and the India Vision 2020 articulates a vision for India's future that throws up a range on interesting possibilities, critical issues and crucial decision points for government and private bodies to ponder over for future action. Many actions for adaptation and mitigation need to be integrated with and are incremental to the ongoing projects to meet our vision and the Millennium Development Goals (MDG). Disaster and climate change are increasingly being considered as a development constraint; hence, mainstreaming them into the development policy is all the more pertinent in the current context.

Centre-State Relations, Policy Issues

The broader and deeper dynamics of Centre-state relations continue to influence both official perceptions and policies towards climate change disaster management. In many, if not most, cases the allocation of responsibility among key agencies within each state and between states and the Central government, is both highly politicised and far from clear.

International Cooperation

Climate change and disasters know no boundaries and will spread across geographical divides. We need to broaden and deepen the nature and scope of the ongoing dialogue to encompass the entire subcontinent. Ironically, climate change offers humanity an opportunity for a quantum leap in sustainable development and in peace-making. If international cooperation—as opposed to competition—is strengthened in response to the threat of climate change and increasing disasters, international stability, governance and development could also benefit. As quoted by the UNDP regarding MDG 8: "Climate change, a global phenomenon, calls for a collective response in the form of global partnerships."[40]

40. UNDP (2007c).

Climate Change: Need for a Single Authority

In India, different ministries feud over turf on climate change. With climate change becoming an important issue both nationally and internationally, several ministries and agencies are now looking to get budgetary allocations. The MoEF, the Planning Commission, the MoST, as well as the recently-constituted office of the Prime Minister's special envoy, are all addressing climate change from their perspectives. The integrated climate change and disaster risk management framework proposed above, needs to be put under one authority/ministry.

REFERENCES

Adger, N. and N. Brooks (2003). "Does global environmental change cause vulnerability to disaster?", in M. Pelling (ed.), *Natural Disasters and Development in a Globalizing World.* London: Routledge.

Anderson, M.B. and P.J. Woodrow (1998). *Rising From the Ashes: Development Strategies in Times of Disaster.* Boulder, CO: Lynne Rienner Publishers.

Benson, C. and E.J. Clay (2001). "Dominica: Natural disasters and economic development in a small island state", *Disaster Risk Management Working Paper Series*. No. 2. Washington, D.C.: World Bank. Available online at *http://www.proventionconsortium.org/files/ dominica.pdf.*

———. (2002). "Bangladesh: Disasters and public finance", *Disaster Risk Management Working Paper Series* No. 6. Washington, D.C.: World Bank. *http://www. proventionconsortium.org/files/ bangladeshfull_doc.pdf*

Blaikie, P., T. Cannon, I. Davies and B. Wisner (1994). *At Risk: Natural Hazards, People's Vulnerability, and Disasters.* First Edition. London: Routledge.

Bodansky, D. (1993). "The United Nations framework convention on climate change: A commentary", *Yale Journal of International Law* 18(2).

Cannon, T. (2000). "Vulnerability analysis and disasters", in D.J. Parker (ed.), *Floods*. London: Routledge.

CRED (2007). *Annual Disaster Statistical Review: Numbers and Trends 2006*. Report by the Centre for Research on the Epidemiology of Disasters (CRED), School of Public Health, Catholic University of Louvain, Brussels, Belgium. May. Available online at *http://www.cred.be and at http://www.em-dat.net/ documents/Annual Disaster Statistical Review 2006.pdf*

DCDC (The Development, Concepts and Doctrine Centre) (2007). *The DCDC Global Strategic Trends Programme 2007-2036.* Third Edition. DCDC, Shrivenham, SWINDON, UK, SN68RF. Available online at *http://www.mod.uk/Defence Internet/AboutDefence/Organisation/ AgenciesOrganisations/DCDC.*

DEFRA/GoI, (2005a). "Indian climate change scenarios for impacts assessments, key sheet 2", Indian Institute of Tropical Meteorology (IITM), Pune, India, in *Investigating the Impacts of Climate Change in India.* Report by Department of Environment, Food and Rural Affairs (DEFRA), UK and Ministry of Environment and Forests (MoEF). Available online at *http://www.defra. gov.uk /environment/climatechange/internat/devcountry/ india2.htm.*

———. (2005b). "Climate change impacts on sea level in India." Key Sheet 4, National Institute of Oceanography, Goa, India, in *Investigating the Impacts of Climate Change in India*. Report by Department of Environment, Food and Rural Affairs (DEFRA), UK and Ministry of

Environment and Forests (MoEF). Available online at *http://www.defra.gov.uk/environment/climatechange/internat/devcountry/india2.htm*

DFID (2005). *Disaster Risk Reduction: A Development Concern.* London: Department for International Development (DFID).

Downing, T.E., A.A. Olsthoorn and Tol R.S.J. (eds.) (1999). *Climate, Change and Risk.* London: Routledge.

Emanuel, K. (2005). "Increasing destructiveness of tropical cyclones over the past 30 years", *Nature* 436: 686-88.

Fankhauser, S. (1998). "The costs of adapting to climate change", *Global Environment Facility (GEF) Working Paper* No. 16. Washington, D.C.: GEF

German Advisory Council on Global Change (WBGU) (2008). *Climate Change as a Security Risk.* UK: Earthscan.

———. (2006). *The Future Oceans-Warming up: Rising High, Turning Sour*. Special Report. Berlin.

GoI (2006). "Crisis management: From despair to hope." *Second Administrative Reforms Commission Third Report.* September.

———. (2004a). *Disaster Management in India: A Status Report*. National Disaster Management Division, Ministry of Home Affairs, Government of India, MHA/GOI/28/06/2004.

———. (2004b). India's Initial National Communication (NATCOM) to the United Nations Framework Convention on Climate Change (UNFCCC). NATCOM 2004. Ministry of Environment and Forests (MoEF), Government of India.

———. (2002). High Powered Committee (HPC) on Disaster Management Report. HPC Report. National Center for Disaster Management, Ministry of Agriculture, Govt of India.

Gosain, A.K., Sandhya Rao *et al.* (2006). "Climate change impact assessment on hydrology of Indian river basins", *Current Science* 90(3): 346-53, 10 Feb.

Guha, Debarati Sapir (2008). *United Nations International Strategy for Disaster Reduction*(UN/ISDR) 2008/01. Disaster Figures for 2007. Press release dated 18 January 2008 prepared by Centre for Research on the Epidemiology of Disasters (CRED). Available online at *http://www.emdat.be/Documents/ conferencePress/ presereleasecred %20jan%202008.pdf*

Handmer, J. (2003). "Adaptive capacity: What does it mean in the context of natural hazards?", in J.B. Smith, R.J.T. Klein and S. Huq (eds.), *Climate Change, Adaptive Capacity, and Development.* London: Imperial College Press.

Hilhorst D. and G. Bankoff (2004). "Mapping vulnerability", in G. Bankoff, G.Frerks and D. Hilhorts (eds.), *Mapping Vulnerability: Disaster, Development and People.* London: Earthscan.

Hulme, M. *et al.* (1999). "Relative impacts of human-induced climate change and natural climate variability", *Nature* 397(6721).

Hultman, N.E., Bozmoski S. Alexander (2006). "The changing face of normal disaster: risk, resilience and natural security in a changing climate". *Journal of International Affairs* 59(2): 25-41, Spring/Summer. Columbia University, New York, USA.

IFRC (International Federation of Red Cross and Red Crescent Societies) (2005). *World Disasters Report 2005: Focus on Information in Disasters*. Geneva.

———. (2003). World Disasters Report 2003. Oxford University Press, Oxford.

IPCC (2007). "Summary for policy makers", in *Climate Change 2007: Synthesis Report.* Cambridge University Press, Cambridge, UK and New York, NY, USA.1-22

———. (2007). "Summary for policy makers", in *Climate Change 2007: The Physical Science Basis.* Contribution of Working Group I to the Fourth Assessment Report of the Intergovernmental Panel on Climate Change [Solomon, S., D. Qin., M. Manning, Z. Chen,

M. Marquis, K.B. Averyt, M. Tignor and H.L Miller (eds)]. Cambridge, UK and New York Cambridge University Press.

———. (2007). "Summary for policy makers", in *Climate Change 2007: Impacts, Adaptation and Vulnerability*. Contribution of Working Group II to the Fourth Assessment Report of the Intergovernmental Panel on Climate Change, M.L. Parry, O.F. Canziani, J.P. Palutikotf, P.J. van der Linden and C.E. Hanson (eds.). Cambridge, UK: Cambridge University Press.

ISDR (International Strategy for Disaster Reduction) (2005). *The Link between Millennium Development Goals (MDGs) and Disaster Risk Reduction*. Available at *http://www.unisdr.org/eng/mdgs-drr/link-mdg-drr.htm*

ISDR/CRED (2007). *2007 Disasters in Numbers*. ISDR/CRED Press Release. Available online at *http://www.emdat.be/ Documents/ConferencePress/ presereleasecred jan2008.pdf*

Murari Lal *et al.* (2001). "Future climate change: Implications for Indian summer monsoon and its variability", *Current Science* 81(9): 1196-1207.

Myers, N. (1993). "Environmental refugees in a globally warmed world", *Bioscience* 43, December.

Olsthoorn, A.A., P.E. van der Werff and J. de Boer (1994). *The Natural Disaster Reduction Community and Climate Change Policy Making: An Inquiry among Participants of the IDNDR World Conference in Yokohama*. Report No. W-94/14. Amsterdam: Institute for Environmental Studies.

Oxfam International (2005). *Paying the Price: Why Rich Countries Must Invest Now in a War on Poverty*. Oxford: Oxfam International.

Parry, M.L., O.F. Canziani, J.P. Palutikof and Co-Authors (2007). *Technical Summary. Climate Change 2007: Impacts, Adaptation and Vulnerability. Contribution of Working Group II to the Fourth Assessment Report of the Intergovernmental Panel on Climate Change.* IPCC, 2007. M.L. Parry, O.F. Canziani, J.P. Palutikotf, P.J. van der Linden and C.E. Hanson (eds.). Cambridge, UK: Cambridge University Press.

Patwardhan, Anand *et al.* (2003). "Impacts of climate change on coastal zones", in P.R. Shukla, Subodh K. Sharma, N.H. Ravindranath, A. Garg and S. Bhattacharya (eds.), *Climate Change and India: Vulnerability Assessment and Adaptation*. Hyderabad: Universities Press (India).

Pelling, M. (2003). "Disaster risk and development planning: The case for integration", *International Development Planning Review* 25(4).

———. (2005). *International data on disaster risk: World Disasters Report* 2005. Geneva: IFRC.

Ribot, J.C. (1996). "Introduction: Climate variability, climate change and vulnerability—Moving forward by looking back", in J.C. Ribot, A.R. Magalhaes and S.S. Panagides (eds.), *Climate Variability: Climate Change and Social Vulnerability in the Semi-Arid Tropics*. Cambridge: Cambridge University Press.

Ribot, J.C., A. Najam, and G. Watson (1996). "Climate variation, vulnerability and sustainable development in the semi-arid tropics", in J.C. Ribot, A.R., Magalhaes and S.S. Panagides (eds.), *Climate Variability: Climate Change and Social Vulnerability in the Semi-Arid Tropics*. Cambridge: Cambridge University Press.

Kumar *et al.,* Rupa (2006). "High-resolution climate change scenarios for India for the 21st century", *Current Science* 90(3), 10 February.

Kumar, Santosh (2006). "Hazard, risk, vulnerability and climate change in India", *Proceedings of the International Conference on Adaptation to Climate Variability and Change*. January 5-7. New Delhi, India. Institute for Social and Environment Transition (ISET), Boulder, Colorado, USA and Winrock International (WII), Delhi, India.

Shukla, P.R. Subodh K. Sharma, N.H Ravindranath, A. Garg and S. Bhattacharya (eds). (2003). *Climate Change and India; Vulnerability Assessment and Adaptation.* Hyderabad, India: Universities Press (India).

———. (2003). "Vulnerability and adaptation: Challenges ahead", in P.R. Shukla, Subodh K. Sharma, N.H. Ravindranath, A. Garg and S. Bhattacharya (eds.), *Climate Change and India; Vulnerability Assessment and Adaptation.* Hyderabad, India: Universities Press (India).

Solomon, S., D. Qin, M. Manning, R.B. Alley, T. Berntsen, N.L. Bindoff, Z. Chen, A Chidthaisong *et al.*, (2007). "Technical summary", in *Climate Change 2007: The Physical Science Basis.* Contribution of Working Group I to the Fourth Assessment Report of the Intergovernmental Panel on Climate Change [Solomon, S.,D. Qin., M. Manning, Z. Chen, M. Marquis, K.B. Averyt, M. Tignor and H.L Miller(eds)]. IPCC. Cambridge University Press, Cambridge, UK and New York, USA.19-71.

UNDHA (1992). *Internationally Agreed Glossary of Basic Terms Related to Disaster Management.* Geneva: United Nations Department of Humanitarian Affairs Publication (DHA). December (DHA/93/36).

UNDP (2007a). *Human Development Report 2007/2008.* New York: PalgraveMacmillan,.

———. (2007b). *Managing Risks of a Changing Climate to Support Development.* Report of the Asia Regional Workshop, 23-26 April 2007, Kathmandu, Nepal, UNDP Regional Center Bangkok, Thailand. *http://regionalcenterbangkok.undp.or.th*

———. (2007c). *Poverty Eradication, MDGs and Climate Change*. Geneva: UNDP, Bureau for Crisis Prevention and Recovery. Available at online *http://www.undp.org/climatechange/adap01.htm*

———. (2004). *Reducing Disaster Risk: A Challenge for Development.* Geneva, UNDP, Bureau for Crisis Prevention and Recovery. Available at *http://www.un.org/special-rep/ohrlls/ldc/Global-Reports/UNDP%20Reducing%20 Disaster %20Risk. pdf*

UN/ISDR (2008/01). *Disaster Figures for 2007.* Press release dated 18 January 2008 prepared by CRED. Available online at *http://www.emdat. be/Documents/ConferencePress/presereleasecred %20jan%202008.pdf*

World Economic Forum (WEF) (2007). *Global Risk 2007*. A Report, in collaboration with Citigroup Marsh and Mclennan Companies (MMC) Swiss Re, Wharton School Risk Center. Available online at *www.weforum.org/en/initiatives/global risk*

4 Climate Change and Migration

P.K. Gautam

Introduction

Two attributes are central to any study of migration and climate change. First is population growth and future projections. As population increases and more areas are threatened by climate change-related impacts, disasters or environmental degradation, more people will get displaced in greater proportion. The second is identifying factors that drive people out like rising sea level, droughts and floods.

The following five factors have been identified as environmental push factors that might lead to migration:

(a) Natural disasters

(b) Development projects that involve change of environment

(c) Progressive evolution of the environment

(d) Industrial accidents

(e) Environmental consequences due to conflicts.[1]

Of these, climate change would influence (a) and (c) significantly, as already highlighted in Chapters 2 and 3.

The challenge with the current state of knowledge is that it is still not possible to separate climate variation with adverse impact of climate change as far as floods and droughts are concerned. Only sea level rise gives a clear indication of areas that will get submerged and how many people may be forced to move out. But rough approximations can be attempted for no regret policies.

This chapter will first trace the process of migration in the recent past. It will analyse population projections till mid-century and discuss

1. Etienne, Piguet (2008). "New Issues in Refugees Research", *Research Paper* No. 153, "Climate Change and Forced Migration", UNHCR, January.

opportunities of urbanisation. It will address sea level rise-induced migration and migration from neighbouring countries and conclude with policy options.

Scale of Migration in the Past

During the partition of India in 1947, there were 14 million people who became migrants, i.e., 8 million Muslims went to Pakistan from India and 6 million Hindus and Sikhs left Pakistan for India. During the war of liberation of Bangladesh in 1971, 10 million migrants from erstwhile East Pakistan fled to India.

Development projects such as dams have caused displacement.[2] Drought-related distress out-migration from mostly rainfed agricultural villages in search of livelihood is well known. As an example, the Bundelkhand region of central India with an annual rainfall of 850 mm, experienced a water crisis due to drought. April 2008 was the fourth consequent year without rains. It faced floods during the 2008 monsoons. The social implications were that seasonal migration increased. Families who had the option to do so, sent their livestock to their relatives' village. Mostly, landless and marginal farmers migrate with or without families for periods ranging from a fortnight to six months or even till the next sowing season. Close to 80 per cent of the villagers had migrated from some villages.[3]

There are pastoral communities of indigenous people who migrate across regions for grazing in the desert, the semi-arid regions and also in the Himalayas.

Flood-related displacement is also now routine. In the northeast, bank erosion of the Brahmaputra leads to loss of land and people get internally displaced. The recent floods in the Kosi river in Bihar in August 2008 has also led people to out-migrate, though temporarily.

People migrating to cities in search of jobs from distress areas or where jobs are not found (as in the Himalayan belt where people migrate to the plains for work) is also a well established trend.

---

2. According to one study the figure for 1951 to 1995 is about 50 million. See Fernandes, Walter and Sanjay Barbora (2008). "Development, Displacement and Right to Life", in Samir Kumar Das (ed), *Blisters on their Feet: Tales of Internally Displaced Persons in India's North East*. New Delhi: Sage Publications. pp.314-336.

3. Kumar, M. Manoj (2008). "Water for All and Always in Bundelkhand Region", *Development Alternatives* 18 (4): 7-11, April.

What would be unique with climate change is that, rather than war being the cause forcing people to migrate, or economic pull or pull factors, climate-related events would force people to move out to safer areas. This in turn will generate tension and in some cases exacerbate conflicts and violence. This is not to say that this has never happened. Many ongoing insurgencies in the northeast are due to illegal migrations. There are projections that indicate that due to the loss of habitat, migration will take place on a large scale. More than the pull factor of greener pastures, climate-related migration will mostly be push factors, though both would work in tandem.

Population Projections

Till 2015: Projections by the United Nations Development Programme (UNDP) for India and neighbouring countries till 2015 is shown at Table 4.1.

Table 4.1

Population Projections for India and Neighbouring Countries

Country	*Total Population (in million)*	
	2005	*2015 (Median Variant Projection)*
Maldives	0.3	0.4
China	1313.0	1388.6
Sri Lanka	19.1	20.0
India	1134.4	1302.5
Myanmar	48.0	52.0
Bhutan	0.6	0.7
Bangladesh	153.3	180.1
Nepal	27.1	32.8
Pakistan	158.1	190.7

Source: *Human Development Report 2007/08*.

Projections till 2026: The population of India is expected to increase from 1,029 million to 1400 million during the period 2001-2026—an increase of 36 per cent in 25 years at the rate of 1.2 per cent annually.[4] As a consequence, the density of population will increase from 313 to 426 persons per square kilometre.

4. Census of India (2001). *Population Projections for India and States 2001-2026* (Revised December 2006). Report of the Technical Group on Population Projections Constituted by the National Commission on Population, Office of the Registrar General and Census Commissioner, India, May 2006. p.x.

Considerable variation in the demographic growth amongst the states has been estimated. The salient features of the projections at the state level are as under:

(a) The state, which is expected to have least growth in the quarter century (2001-2026) is Tamil Nadu (15 per cent), followed by Kerala (17 per cent). In contrast, Delhi will have the highest projected growth of 102 per cent during 2001-2026. States, which will have projected growths in the range of 20-30 per cent are Himachal Pradesh, Punjab, West Bengal, Orissa, Andhra Pradesh and Karnataka. The population in the states of Haryana, Rajasthan, Uttar Pradesh and Madhya Pradesh is projected to increase by 40-50 per cent during 2001-2026, which is above the national average of 36 per cent. The population of Uttar Pradesh is expected to be highest among all the states of the country, at almost 249 million in 2026.

(b) Of the projected increase in population of 371 million in India during 2001-2026, 187 million is likely to occur in the seven states of

Figure 4.1

Percentage Share of States in Total Projected Population Increase during 2001-2026

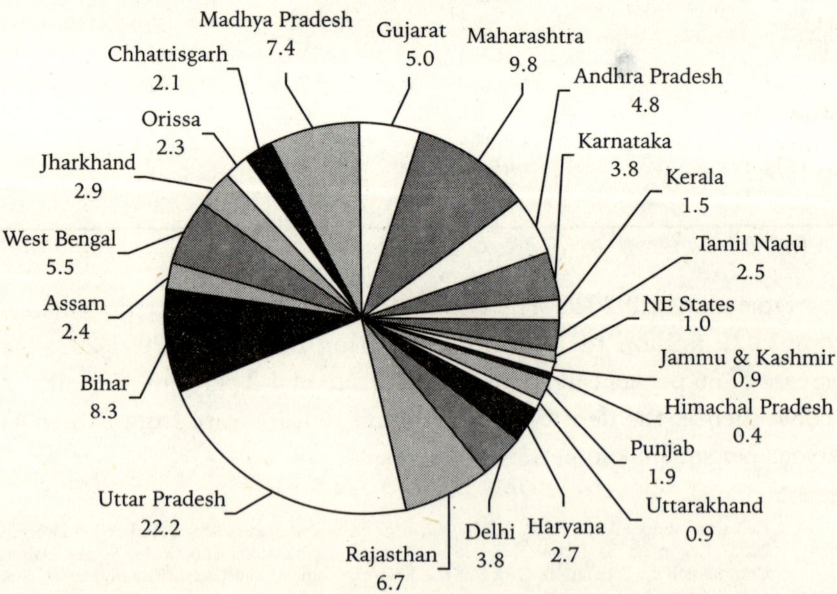

Bihar, Chhattisgarh, Jharkhand, Madhya Pradesh, Rajasthan, Uttar Pradesh and Uttarakhand (the BIMARU states, before division). Thus, nearly 50 per cent of India's demographic growth during this period of 25 years, is projected to take place in these seven states. Twenty-two per cent of the total population increase in India of 371 million during 2001-2026 is anticipated to occur in Uttar Pradesh alone. The population in these seven states together is expected to grow at 1.5 per cent per annum during 2001-2026. In contrast, the contribution of the four southern states, namely, Andhra Pradesh, Karnataka, Kerala and Tamil Nadu, to the total increase in population of the country during 2001-2026 is expected to be 47 million—13 per cent of the total demographic growth of the country. The population in these four states together is expected to grow at 0.8 per cent per annum during 2001-2026. The contribution of some of the other states to the total increase in population size during 2001-2026 are expected to be: Maharashtra—9.8 per cent, West Bengal—5.5 per cent, Gujarat—5 per cent and Delhi—3.8 per cent. These four states together thus would contribute about 24 per cent of the total increase in population during 2001-2026. The contribution of the remaining states and union territories to the total increase in population size during 2001-2026 is expected to be 13 per cent (Figure 4.1).

Projections up to Mid-Century: The projections of population up to the mid-21st century are as under:

(a) NATCOM: 1.6 billion (160 crore) by 2050 (NATCOM: 83). This figure is given in context of greater demand for food.

(b) IPCC's Working Group's Chapter 10 on South Asia:

The majority of the region's population growth is forecast in South Asia, which expects to add 570 million people in India, 200 million in Pakistan and 130 million in Bangladesh over the next 50 years.[5]

(c) Population projections as given in provisional population totals of Census of India 2001 also allude to this:

---

5. Cruz, R.V., H. Harasawa, M. Lal, S. Wu, Y. Anokhin, B. Punsalmaa, Y. Honda, M. Jafari, C.Li and Huu Ninh (2007). "Asia Climate Change 2007: Impacts, Adaptation and Vulnerability", in M.L. Parry, O.F. Canziani, J.P. Palutikof, P.J. van der Linden and C.E. Hanson (eds.), *Contribution of Working Group II to the Fourth Assessment Report of the Intergovernmental Panel on Climate Change*. Cambridge, UK: Cambridge University Press. p. 488.

"It is also important to take note that the perils of long term projections notwithstanding, it is felt that based on the medium term projection, India is all set to overtake the population of China in possibly less than fifty years and continue to add to its population even beyond 2050" (Long-range World Population Projections: Based on the 1998 Revision, UN, New York, 2000).

To visualise the problem of pure high population *per se*, we can conclude the following population projections by 2050 (Figure 4.2):

(i) India 160 crore

(ii) Pakistan 35.8 crore

(iii) Bangladesh 28.33 crore

Figure 4.2

Population Projections

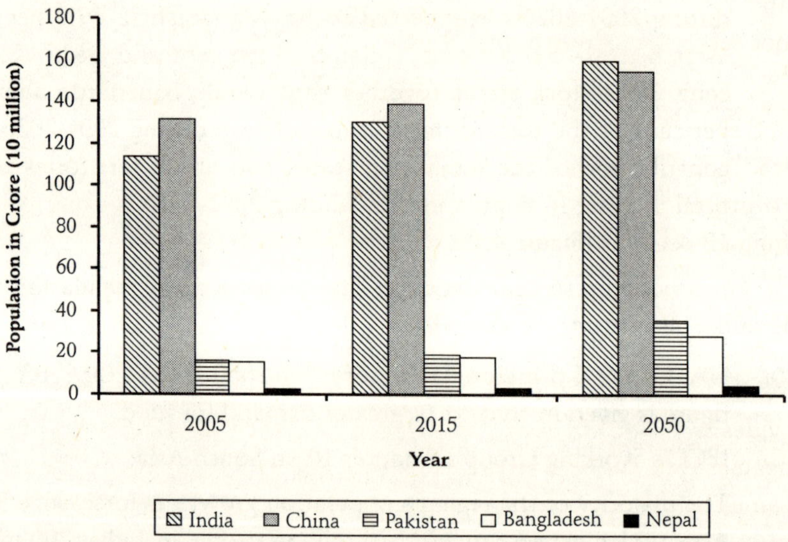

Sources: *Human Development Report 2007/08*, NATCOM, IPCC's Working Group Chapter 10 on South Asia, Census of India 2001.

Urban Growth

"In Asia, migration accounts for 64 per cent of urban growth. Climate–related disruptions of human population and consequent migration can be expected over coming decades. Periods when precipitation shortfalls coincide with adverse economic conditions for farmers would be those most likely to lead to sudden spikes in rural-to-urban migration levels in China and India. Climatic changes in Pakistan and Bangladesh would likely exacerbate present environmental conditions that give rise to land degradation, shortfall in food production, rural poverty and urban unrest. Circular migration patterns, such as those punctuated by shocks of migration following extreme weather events, could be expected."[6]

Urban Population

The urban Indian population was 28 per cent in 2001 and is expected to increase to 33 per cent by 2026. Populations of mega cities by 2015 are projected to be: Mumbai 2.74 crore, Calcutta and Delhi 1.76 crore and Hyderabad 1.07 crore. By 2047, half of India will be urban.

With such concentration of population, climate-related disasters focused on a small geographical area may lead to more human casualties and suffering due to high population densities. Further, urban centres do not seem to be geared up to receive the migrants.

Nearly one-third of the urban growth will manifest itself in slums and shanty towns. If the aspirations of people are not met, then conditions leading to tensions and conflicts in urban areas by spread of Naxalism to urban areas, or a sharper divide between the "sons of soil" with the migrant "outsiders", or social tensions due to an unbearable rich and poor divide who live 'cheek by jowl' could be highly expected. Already some political forces have demonstrated that "outsiders" are unwelcome in states and must go back (as in Assam in 2007 and in Mumbai in 2008). While in a tolerant liberal society no major conflicts have erupted, recent widespread violent reactions from places such as Mumbai or Guwahati against the perceived intruders portent potential large-scale conflict. These same trends will continue and grow with rising population and climate-led migration, over and above the "normal" movement of people to greener pastures, will further exacerbate the scale and tempo.

Migration Due to Rise in Sea Level

The vulnerability of India's coast to rising sea levels has been analysed in NATCOM.

6. Ibid.

NATCOM indicates that a one metre sea level rise is projected to displace approximately 7.1 million people in India and about 5,764 square kilometres of land area will be lost, along with 4,200 kilometres of roads. It further amplifies that the extent of vulnerability depends not just on the physical exposure to the rise in sea level and affected population but also on the extent of economic activity of the areas and capacity to cope with impacts. However, it will be an alarmist outlook to work on projections of one metre rise *ab initio* leading to forced displacement. As analysed in Chapter 2, under most conservative scenarios, sea level could be expected to be about 40 cm higher than today by the end of the 21st century. It is projected that this would increase the number of people annually in the coastal population, from 13 million to 94 million. Almost 60 per cent of this increase will occur in South Asia (along the coast from Pakistan, through India, Sri Lanka and Bangladesh).[7]

Recent reports such as Greenpeace's Blue Alert: Climate Migrants in South Asia (March 2008) by Professor Sudhir Chella Rajan from IIT Madras treat IPCC estimates as conservative. It draws heavily from the findings of the report of the German Advisory Council on Global Change (WBGU). According to their findings of the business as usual (BAU) project, sea level rise is expected to be to the tune of 1 m to 5 m by 2100 and not a maximum of 59 cm by the end of the century, as in the IPCC estimates. The report includes factors—besides thermal expansion—like melting/break-up of ice sheets in Greenland and Antarctica. The report shows that if global temperature rises by about 4-5 degree celsius in the course of the century, about 125 million migrants (including 75 million from Bangladesh) may be rendered homeless by the end of the century. The scenario takes note of impacts of climate change, including sea level rise and drought, associated with shrinking water supplies and monsoon variability.

Risk of Migrant Population from Neighbouring Countries

It will be prudent to take into account the inflow of migrants from Nepal and Bangladesh based on current trends. Tim Dyson and Pravin Visaria quoting UN data for population of Bangladesh and Nepal as 21.1 and 4.9 crore respectively by 2025, had projected that the volume

7. Ibid.: 484.

of migration from Bangladesh and Nepal would increase.[8] Even without climate change-related migration, migrants from Nepal and Bangladesh do present a security challenge to India which is likely to exacerbate with climate-related migrations.

With projected population growth, migrants to India from Nepal and Bangladesh are only bound to increase. Figures of nearly 20 million illegal Bangladeshi migrants and unofficial estimates indicate that about 6 million Nepali citizens in India are residing in India without any climate-related reasons.[9] But both population growth and ecological degradation due to climate change will tend to encourage flow towards India. It may never be possible to establish how much would be due to climate change or rise in sea level, Himalayan degradation or loss of fertility and other reasons. However, with more extremities such as droughts and floods and their impact on productivity, migrants may be the most direct cause of tensions in the near future.

Conclusion

Together with population growth, it can be concluded that within India migration would increase as under:

(a) Traditional rainfed rural to urban migration will now include additional distress areas as a result of climate change-induced droughts and floods.

(b) From hills to plains as in the past.

(c) From BIMARU states as there will not be enough jobs/livelihood.

(d) From coastal regions to the hinterland due to the rise in sea levels:

- (i) Rural migrants from the hinterland may return to their villages and towns to get adjusted in their original environment.
- (ii) Migration from coastal urban to urban in hinterland and beyond.

---

8. Dyson, Tim and Pravin Visaria (2004). "Migration and Urbanization", in Tim Dyson, Robert Cassen and Leela Visaria (eds.), *Twenty-First Century India: Population, Economy, Human Development and the Environment*. Chapter 6. New Delhi: Oxford University Press.

9. According to Dr. Pushpita Das, 2.2 million Nepali citizens reside in India. Unofficial estimates are 6 million. See Das, Pushpita (2008). "Towards a Regulated Indo-Nepal Border", *Strategic Analysis* 32 (5): 885, September.

(e) As cities on the coast would no longer be attractive, hinterland cities may get more migrants than in the normal circumstances.

(f) With the slowdown of the world economy and also overcoming of the Dutch disease, West Asia may not be able to absorb so many Indian workers as in the past. Kerala for instance, may not be so lucky to have a sizeable number of migrant workers in the Gulf countries in the future. Domestic pressures on joblessness will increase. This has been included as a trend as it is not linked to climate change events. It highlights forces that would create social pressures.

Due to the difficulties in sifting out data for analysing pure contribution of climate change to the scale of migration, there would, in all likelihood, be a tendency to evoke the narrative of climate change and cover up the lack of governance and routine coping capacities, and place the blame on external climate changes.

As far as cross-border migration is concerned, flows from Nepal and Bangladesh will increase in proportion to population growth in those countries. Even from India, some overflow from the bordering region of Bihar and UP to Madesh in Nepal may take place.

The case of Pakistan is crucial. If water-related stress and soil degradation continues, then quite a sizeable population could migrate to India, as the trend now is for more people-to-people contact. Fences, in any case, are also porous.

All examples are prone to generate tension and may not bring any happiness or prosperity. There are chances that the fascist concept of lebensraum may enter the discourse.

Policy

For intra-state migration, a national debate must be initiated to place the problem in proper perspective due to all factors including climate change. Within India, democratic institutions need to be strengthened. While all measures need to be initiated to negate the "sons of soil" theory, allowing free movement of people without necessary checks and balances could lead to high conflict situations.

Counter-migration and urbanisation strategies need to be put into place by creating diverse livelihood options *in situ*. The Gandhian model needs to be augmented with the PURA (providing urban amenities in rural areas) model of the former President of India Shri Abdul Kalam. Also, urban areas must now include about 10 per cent extra population figures for future planning to absorb likely migrants.

Priority should be given to areas impacted by rising sea levels and saltwater intrusion. Adaptation such as provisions for sea level rise must now include measures beyond dykes and sea walls. Further, growth/spread of coastal cities at risk must not be allowed. Military and the central police forces need to be trained to handle migration-like situations as during the partition, with a humanitarian touch. Population policy must be revised to include the role of education in bringing down the total fertility ratio.

The foreseeable problem of likely influx of climate migrants from neighbouring states—especially Bangladesh and Nepal—needs to be discussed jointly with the countries to device suitable strategies to avoid escalation of inter-state conflict. Appropriate initiatives within SAARC also need to be commenced to deliberate on the challenge.

5 Energy and Climate Change

Shebonti Ray Dadwal

The linkages between energy and climate change are now well established as is the fact that the build-up in emissions of carbon and other greenhouse gases (GHG) in the atmosphere are a result of burning of fossil fuels. The dilemma, however, particularly for developing countries like India, is whether the need for increased use of energy for economic development is more important than limiting or mitigating carbon emissions from fuel burning. For India, where power shortages leave more than 400 million people without access to electricity, particularly in poor rural areas, the need to increase the power generation capability several-fold is crucial, as are adequate supplies of oil for the transport sector.

Politics of Energy

The oil—and indeed the energy market—is undergoing a transformation as demand shifts from the traditional Western developed countries to the Asian developing countries. As Asian markets grow in importance for energy producers, they are no longer as dependent on the traditional developed countries. Moreover, as production of oil in the developed countries continues to drop, West Asian and Former Soviet republics are gaining in stature as the most important producers. This has allowed many of the producing countries to resort to resource nationalism, and some of them (mainly Russia and the Latin American countries) have adopted a somewhat confrontationist attitude with the US and the EU.

At the same time, growing demand for energy—both global and regional—saw the prices of almost all energy resources, particularly hydrocarbons, reach unprecedented levels. In turn, this is affecting the global economy, which has shown a considerable slowdown.

As one of the largest growing energy consumers, India along with China, is being subjected to pressure by the developed countries to reduce its energy consumption. At a seminar on trade, finance and climate in Brussels on September 15, 2008, Ditte Juul-Jorgensen of the European Commission's trade directorate made a reference to import restrictions or border tariffs that may be imposed by developed states to prevent carbon leakage[1] (see the section on energy, climate change and international trade later in the Chapter).

Despite the fact that their energy consumption, and hence their carbon emission levels are far less than that of the developed countries in real terms (India's oil consumption is around 2.8 million barrels a day (mbd), compared to the US's 21 mbd), the developing countries are being held responsible for resource depletion, higher energy prices and growing emissions.

India's Projected Emissions

As one of the fastest growing economies in the world, India is being targeted for its growing energy consumption level and consequently GHG emissions. But statistics indicate that between 1990 and 2005, emission levels in the country were very low. The International Energy Agency's (IEA) World Energy Outlook (WEO) 2007, in which three scenarios—namely, the reference case scenario, the alternative policy scenario and the high growth scenario—have been projected, India's historical share of cumulative emissions between 1990 and 2005 amounted to only 4.6 per cent. China's is 8 per cent, the US and the EU countries accounted for more than half of the cumulative emissions over the same period. (See Table 5.1).

However, according to the IEA WEO 2007, India's emissions are projected to rise to 14 per cent in the reference scenario (RS), 17 per cent in the alternative scenario (AS) and 15 per cent in the high growth scenario (HGS). China's on the other hand, is projected to increase to 42 per cent in the RS, 52 per cent in the AS and 49 per cent in the HGS. (Table 5.2). One projection as total GHG emission by country/region is given at Table 5.3.

India is also one of the world's lowest-cost producers of aluminum and steel, and while its GDP has grown at an annual average of 8 per cent, its energy consumption has risen by only 4 per cent. In the outlook period

1. "Experts warn EU of climate change 'trade war'", *EurActiv.com*, September 16, 2008, at *www.euractiv.com/en/climate-change/experts-warn-eu-climate-change-trade-war/article-175426?Ref=RSS*

too, i.e., till 2030, India's per capita emissions are projected to rise to only 4 per cent in the RS (and around 6-8 per cent in the HGS). China's emissions, on the other hand, are projected to rise to 16 per cent. But the developed world usually lumps the two countries together for energy consumption and carbon emissions. As the above figures indicate, it would, however, be fallacious to place India and China in the same category.

Table 5.1

Top 20 Carbon Emitters

Countries	Total	Emissions (mt CO_2)	Growth Rate (%)	Share of World Total (%)		CO_2 Emission Per Capita (+CO_2)	
	1990	2004	1990-2004	1990	2004	1990	2004
USA	4,818	6,046	25	21.2	20.9	19.3	20.6
China	2,399	5,007	109	10.6	17.3	2.1	3.8
Russia	1984	1524	-23	8.7	5.3	13.4	10.6
India*	682	1342	97	3.0	4.6	0.8	1.2
Japan	1071	1257	17	4.7	4.4	8.7	9.9
Germany	980	808	-18	4.3	2.8	12.3	9.8
Canada	416	639	54	1.8	2.2	15.0	20.0
UK	579	587	1	2.6	2.0	10.0	9.8
S. Korea	241	465	93	1.1	1.6	5.6	9.7
Italy	390	450	15	1.7	1.6	6.9	7.8
Mexico	413	438	6	1.8	1.5	5.0	4.2
S. Africa	332	437	32	1.5	1.5	9.1	9.8
Iran	218	433	99	1.0	1.5	4.0	6.4
Indonesia	214	378	77	0.9	1.3	1.2	1.7
France	364	373	3	1.6	1.3	6.4	6.0
Brazil	210	332	58	0.9	1.1	1.4	1.8
Spain	212	330	56	0.9	1.1	5.5	7.6
Ukraine	600	330	-45	2.6	1.1	11.5	7.0
Australia	278	327	17	1.2	1.1	16.3	16.2
Saudi Arabia	255	308	21	1.1	1.1	15.9	13.6

Note: * According to the *IEA World Energy Outlook*, Japan is the fourth largest emitter followed by India.

Source: UNDP, *Human Development Report 2008-09*.

Developed countries' political rhetoric apart, there is nonetheless no doubt that as India is and will continue to be hugely dependent on coal for the power sector and oil for the transport sector, it will have a deleterious impact on its environment. Policymakers are awakening to the problems that climate change could create for the country, and are taking steps to deal with the same. Prime Minister Manmohan Singh has

Table 5.2

Projections of CO_2 Emissions by Region (in Billion Tonnes)

Country	*Reference Scenario*			*Alternative Scenario*		*High Growth Scenario*	
	2005	*2015*	*2030*	*2015*	*2030*	*2015*	*2030*
OECD	12.8	14.1	15.1	13.2	12.5	13.9	14.6
N. America	6.7	7.5	8.3	7.2	7.1	7.5	8.1
USA	5.8	6.4	6.9	6.2	6.0	6.3	6.7
Europe	4.0	4.2	4.5	3.8	3.5	4.3	4.4
EU	3.9	4.0	4.2	3.6	3.2	4.1	4.2
Pacific	2.1	2.3	2.3	2.2	1.9	2.2	2.1
Transition Economies	2.5	3.0	3.2	2.9	2.8	3.0	3.2
Russia	1.5	1.8	2.0	1.7	1.7	1.8	2.0
Developing Countries	10.7	16.4	22.9	15.2	17.9	17.4	26.3
China	5.1	8.6	11.4	8.1	8.9	9.5	14.1
India	1.1	1.8	3.3	1.6	2.4	1.9	2.6
Other Asia	1.4	2.0	2.7	1.8	2.1	2.0	2.6
West Asia	1.2	1.8	2.5	1.7	2.0	1.8	2.7
Africa	0.8	1.0	1.4	0.9	1.1	1.0	1.3
L. America	0.9	1.2	1.6	1.1	1.3	1.2	1.6
World	26.6	34.1	41.9	31.9	33.9	34.9	44.8

Source: IEA, *WEO 2007*.

Table 5.3

Shares of Total GHG Emissions by Country/Region (%)

Country/Region	*2005*	*2030*	*2100*
USA	18.7	11.2	5.5
China	18.7	33.8	23.2
EU	12.9	7.4	3.5
FSU	8.3	6.4	3.8
India	4.9	8.7	19.3
OPEC	4.7	5.3	7.7
Japan	3.5	1.8	0.6
Canada	2.0	1.3	1.0
Indonesia	1.8	1.7	2.4
Other Southeast Asia + S. Korea	3.0	2.2	2.8
South Africa	1.3	1.3	1.3
Rest of World	18.6	17.9	27.8
World	100.0	100.0	100.0

Source: Garnaut Climate Change Review, Chapter 4.

formed a committee to cooperate on the issue. The National Action Plan on Climate Change that was recently formulated, focuses among other things, on boosting energy efficiency, increasing the share of solar energy, improving seed types and encouraging forestation. Many of these are built upon existing policies and the government claims that over 2 per cent of its GDP is being spent on coping with climate-induced problems.

Moreover, at the June 2007 G8 summit the Prime Minister pledged that India's per capita carbon emissions would never exceed those of the developed countries.[2] Nevertheless, India is opposed to setting targets, and certainly the proposal that GHG emissions have to be cut by 25-40 per cent by 2020, is not acceptable to India. In this, it is supported by other developing countries, and the Bali draft report has now been watered down to include the provision that the developing countries will undertake "measurable, reportable and verifiable" efforts to cut their emissions. However, according to Prodipto Ghosh, Secretary, Ministry of Environment and Forests, and one of the principal interlocutors on the country's climate change negotiations, India's existing energy policy would cut its GHG emissions by more than 25 per cent by 2020 but warned that pressure to set mandatory targets to curb global emissions would impact on economic growth.[3]

Steps Taken to Arrest Emissions

Despite India's reluctance to adhere to demands on setting time-bound targets for emissions cuts, the country has a history of addressing concerns related to environmental and climate change. For example, the Energy Conservation Act (ECA), 2001 came into force with effect from 1 March 2002. Some of the salient features and implementation aspects of the ECA are:

- Setting up of a Bureau of Energy Efficiency (BEE) whose mission would be to institutionalise and strengthen the delivery mechanism for energy efficiency services and provide leadership to energy efficiency in all sectors of the economy.

---

2. Ramachandran, R. (2007). "Class injustice", *Frontline* 24 (24), December 08-21.
3. "India says emissions to fall 25% by 2020", Reuters, May 28, 2007, at *www.reuters.com/article/environmentNews/idUSDEL23553920070528*

- Its prime objective would be to reduce energy intensity in the Indian economy.
- To set up an Energy Conservation Fund at the Centre to develop the delivery mechanisms for large-scale adoption of energy efficiency services such as performance contracting and promotion of energy service companies.
- The fund is expected to give a thrust to R&D and a demonstration in order to boost market penetration of efficient equipment and appliances as well as support the creation of facilities for testing and development and to promote consumer awareness.
- Evolving minimum energy consumption and performance standards for notified equipment and appliances.
- Prohibit manufacture, sale and import of such equipment which does not conform to the standards.
- Introduce a labelling scheme for notified appliances.
- Disseminate information to consumers.[4]

Moreover, the state governments have been asked to amend the energy conservation building codes to suit regional and local climatic conditions. According to the BEE statistics, steps taken to implement efficiency norms in the power sector have resulted in energy savings by 123 participating units under the Ministry of Power's Energy Conservation Award Scheme between 1999 and 2007 of around 1,420 MW, 51.61 lakh tonnes of coal and 1,91,979 cubic metres of gas.

Moreover, through the carbon trading projects under the Clean Development Mechanism (CDM), more than 5 million tonnes of carbon dioxide will be avoided by 2012 by cutting back 10 per cent of the country's GHG emissions every year. The government has cleared more than 1,000 projects for carbon trading, the highest in the world, followed by China, attracting investments worth Rs. 119,662 crore in these projects.[5] However, though India leads in terms of the number of CDM projects, China has a 10 per cent lead over India in terms of number of Certified Emissions Reduction (CER) generation potential from registered projects.

---

4. "Conservation Activated", Vol.1, Issue 1, Octtober-December 2001, *http://bee-india.nic.in/sidelinks/energize/energise1/lead.htm*
5. Times of India (2008). "India will avoid 5 million tonnes of CO_2 emissions by 2012", *The Times of India*, September 8.

This is because the share size of China's economy is larger. The Indian CDM sector, on the other hand, mostly comprises of small to medium-sized efficiency projects requiring low levels of upfront investment.

India is now trying to improve its efforts and the results seem encouraging. In July 2008, the Indian Venture Capital Firm (IVCF) launched the $80 million Green India Venture Fund, which will invest the majority of its capital in projects earning CER credits. Another fund, the New York-based Green Ventures International, launched a $300 million India Carbon Fund earlier this year.

Moreover, the National Action Plan on Climate Change[6] is focusing on solar energy, calling for increases of 1000 MW per year of photovoltaic and solar thermal power generation, respectively.

The second factor that may encourage developing countries like India to become greener is the growth of indigenous alternative energy companies. India has a robust renewable energy programme and plans to increase the share of renewables in its overall energy mix. Indian companies too have made progress in developing renewable technology, and Suzlon Energy is one of the world's five biggest makers of wind turbines.

India is also participating in the IEA projects such as Greenhouse Gases Research and Development, Clean Coal Centre and Demand Side Management, as well as the $1 billion FutureGen project being developed by the US Department of Energy, which aims to create a zero-emission coal-fired plant with carbon capture technology.[7]

Efficiency in use of electricity is one of the most effective as well as cost-effective ways to mitigate emissions. According to some statistics, residential and commercial buildings account for 60 per cent of electricity demand. Hence, cutting the use of electricity by half could lead to substantial saving of carbon. Tata Power is setting up a 4,000 MW coal-based Ultra Mega Power Project at Mundra in Gujarat based on supercritical technology, that has a relatively lower carbon footprint than existing coal plants, and is considered the option for India under the current and future prospects. The International Finance Corp. (IFC) is sponsoring the project.

6. "National Action Plan on Climate Change", PM's Council on Climate Change, Government of India, at *http://pmindia.gov.in/Pg01-52.pdf.*

7. "China and India Insights", *World Energy Outlook 2007*, International Energy Agency, pp. 231-232.

The GHG emissions per kilowatt hour of energy generated by the plant will be about 750 grams of CO_2/kWh, as compared to India's national average of 1,259 grams of CO_2/kWh for coal-based power plants. The world average is 919 grams of CO_2/ kWh, while the average for OECD countries is 888 grams of CO_2/ kWh (2005 figures). As compared to any other sub-critical power plant in India, this project will avoid burning 1.7 million tonnes of coal per year, thus averting carbon emissions of 3.6 million tonnes per year, according to the IFC's documents. Environmentally, it is the cleanest commercially viable pulverised coal technology. The advantage of using ultra supercritical technology over other types of pulverised coal technologies is that it takes less energy to convert the water used in the power generating process to steam. This means that less fuel needs to be burned to generate the same amount of power. When compared to older technologies, the ultra supercritical power plants operate at increased efficiency and as such, have considerably lower fuel costs.

For example, the average efficiency of an older coal plant is around 36-38 per cent while a newer supercritical plant can achieve an overall efficiency in the range of 43-45 per cent and the most advanced ultra supercritical plant can achieve up to 50 per cent efficiency, according to information provided on various websites.

These increases in plant efficiency can reduce CO_2 emissions by a ratio of 2 to 1 (i.e., a one percentage point increase in efficiency reduces emissions by around 2 per cent).

Though development of alternative energy, if made popular and commercially viable, would go a long way in mitigation, they can at best co-exist with fossil fuels and act as supplementary fuels at least for the foreseeable future. As long as there are abundant reserves of oil, gas and coal, they will continue to be the most widely used fuels, and certainly by the developing countries, both for their cost as well as the fact that it would entail a restructuring of their economies and technology regimes. Even if oil production does peak in the near future, coal and its derivatives could tide the earth over for more than a century.

Technology Transfer

The focus of the debate today is how to nurture further economic growth without adding to emission levels. All countries, developed and developing, want to avoid cutbacks or restrictive energy policies which would undermine their growth. For this, technological innovation is the best way out of this dilemma.

Technological solutions are needed for the challenges of both mitigation and adaptation. But in dealing with energy-related issues pertaining to climate change, mitigation technologies are more important. These include development of renewable energy sources such as biofuels, biomass, wind, solar and hydropower, increased efficiency in energy consumption, including in industry as well as infrastructure and buildings, and emerging technologies which aim to capture carbon out of the atmosphere and store it.

However, the efficacy of mitigation solutions based on technological innovations will depend on their being deployed on a global scale and not by only a handful of developed countries. In fact, if one takes the IEA estimates into account that 60 per cent of GHG emissions will come from the developing countries by 2030, it is these countries that will need to have access to mitigation technologies. For this, technology transfer from developed to developing countries as well as between developing countries will be required. However, given that many of these technologies are still under development and are extremely expensive, gaining access to them at an affordable rate will be difficult.

While the developed world is pressurising countries like India to adopt cleaner energy technologies, often they are not willing to allow easy access to the same. Today, nuclear power and Clean Coal Technology (CCT) involves Carbon Capture and Storage (CCS) facilities in coal plants. As India is highly dependent on coal for its power sector, it will have to look at means to enhance power generation with minimal emissions. The options available in coal plants include supercritical plants using pulverised coal steam cycle and the newer Integrated Gasification Combined Cycle (IGCC), which can include carbon sequestration technology but is more expensive than coal steam-electric power.

The cost of CCS depends on the type of plants, distance to the storage site, etc. However, according to a study, the incremental cost of CCS would be around Re. 1/kWh over and above the generation cost of Rs. 2.25/kWh for a coal IGCC plant. The study also says that applying CCS to new coal plants would be cost efficient only if the cost of emitting CO_2 was at least Rs. 1,160 to Rs. 1,400 ($25-30) per metric tonne.[8]

Supercritical coal combustion plants operate at increasingly higher temperatures and pressures and therefore, achieve higher efficiencies than conventional coal units as well as significant CO_2 reductions.

While nuclear energy is an option that India is pursuing with vigour, CCT technology has limited applicability in the case of India, due to the uncertainty surrounding the issue of carbon sequestration as well as high cost of installing plants with carbon capture technology. A 2005 UK Parliamentary Study estimated that building a 500 MW IGCC plant would cost $686 million without CCS and $930 million with it. Also, even if ongoing trials succeed in proving that GHGs can be safely sequestered for 10-20 years, the question of whether they can be safely sequestered for 50 years or more will remain.

Energy, Climate Change and International Trade

During the climate conference, COP13 in Bali, the Indonesian government took the initiative to hold an informal meeting between the ministers of trade on the possibilities of evolving a trade policy that would contribute to the fight against climate change. The talks dealt, among other things, with the joint EU-US proposal of December 3, 2007, to reduce tariffs on environment-friendly goods and services and on transfer of technology to developing countries. It was decided after the Bali meet that there was need for further studies of the linkages between trade and climate.

Currently, much of the problems associated with emissions mitigation, particularly in the developed world, is centred on how to compel domestic industries to reduce emissions when competing industries in the developing countries may be subjected to less stringent emission rules. Many are convinced that the answer lies in using trade regulations as a mechanism to address such concerns as well as a lever to attempt to reduce GHG emissions in a globally coordinated fashion. As Pascal Lamy said, "The relationship between international trade and ... the WTO and climate

8. Hawkins, David G., Daniel A Lashof and Rovert H Williams (2006). "What to do about Coal?", *Scientific American*. September. pp.36-43.

change would be best defined by a consensual, international agreement on climate change that successfully embraces all major polluters."[9] The developing countries, however, feel this could be a means to put pressure on them to have common binding rules on emission cuts.

Developed countries, particularly the EU, are concerned that without a truly international climate change deal in place, their heavy industries would not be able to compete on a global scale with producers operating in countries where pollution is cheaper. According to them, if they move faster to adopt environment policies than their competitors, it could encourage energy or carbon-intensive industries to shift their production outside of Europe to avoid the added costs of an Emissions Trading Scheme (ETS). In other words, global carbon levels would not be reduced; they would simply leak outside of the EU system. Second, energy or carbon-intensive industries that continue in Europe could face additional costs linked to emissions trading credits, which they find difficult to pass on to customers and hence erode their international competitiveness. In fact, EU leaders such as French President Nicolas Sarkozy and European Commission President Jose Manuel Barroso have repeatedly referred to the possibility of imposing a carbon tax or allowance requirement on imports from countries that are not in compliance with the Kyoto Protocol emission reduction requirements.[10]

These suggestions have drawn strong criticism not only from the developing countries but also from the US trade officials, who warn that these proposals could facilitate WTO-inconsistent trade protectionism under the guise of environmental protection. But at the same time, the US Congress is moving to create a system of trade sanctions that would levy heavy taxes on imports from other major GHG emitters.

On the other hand, proponents of free trade argue that import barriers would only lead to a diversion of 'dirty' exports and provide no incentive for developing states to obtain clean technologies and/or to push through industrial emission reductions.

---

9. "Nations bicker in Bali over 'green' goods trade", Reuters, December 9, 2007, at *www.reuters.com/article/latestCrisis/idUSSP210166*

10. Bernd G. Janzen, Akin Gump Strauss Hauer and Feld L.L.P., "Trade Law and Climate Change: Convergence or Conflict?", at *www.metrocorpcounsel.com/current.php?artType=view&artMonth=March&artYear=2008&EntryNo=8014*

Currently, there appears to be little consensus on the issue. In fact, the failure of the Doha Round in the WTO negotiations has raised questions about whether or not a global climate change deal can be effective if the multilateral trading regime under the WTO itself continues to be in doubt. While Lamy argues that a global climate change deal should precede a multilateral trade regime so that WTO members can adjust their trading stances, others disagree on the ground that existing WTO rules and mechanisms could be strengthened to provide positive inputs into climate change negotiations.[11]

11. No. 1.

6 Economics and Adaptation to Climate Change

Purnamita Dasgupta

Some Relevant Aspects

Climate Change Science, Human Security and Vulnerability

There is growing concern that an economic growth strategy which seeks to be both inclusive and sustainable should integrate climate change effects in order to ensure that the long-term consequences of development activities are not detrimental to human well-being. Given the likely adverse impacts, the unpredictability of these impacts adds several dimensions to the analysis of economic implications of the consequences of climate change. It also bears implications for maintaining peace and human security, both within the Indian economy and in its relations with the rest of the world. In the following paragraphs, the sources and issues likely to be of concern in the decades to come are briefly discussed.

The current initiatives for tackling climate change impacts are at a nascent stage. There is ample scope for research and development which focuses on the priorities of the Indian economy. Security-related issues would be one such area where research needs to be initiated. Discussions on ensuring security in terms of access to scarce resources such as water and energy, and in achieving food security have been deliberated upon. However, the threats to internal peace and stability that may arise from the impact of climate change or the additional demands that these may make on resources devoted to maintaining peace and minimising conflicts, also need to be studied in the specific context of the Indian economy. Three important factors that affect human security both within the economy and in terms of relations with the rest of the world will be trade and trade-related barriers, migration and the "common concerns" of sharing available food, water and energy resources. Such concerns may also inform strategies for coping with climate change as we take note of the regional and global interdependencies of climate change concerns for South Asia.

In terms of scientific evidence of climate change, there is reasonable consensus on the implications for South Asia. Initial findings from NATCOM studies too, point out that several regions in India are vulnerable, as already analysed in Chapters 2 and 3. Vulnerability in poverty literature refers to the possibility that consumption of a household or an individual can fall below some socially accepted norms. Therefore, going by the standard notions of vulnerability in terms of poverty, India cannot afford to ignore the economic implications of climate change. Consider for instance, the case of food security. Here, vulnerability is determined in terms of the options available to households to make a living after taking due note of the risks that they face in making a living. Climate change can add to these risks and also affect the capacity to handle these risks. To quote from the IPCC, vulnerability is defined as "the degree to which a system is susceptible to, or unable to cope with, adverse effects of climate change, including climate variability and extremes."

It is evident that climate change impacts will have implications for major sectors in the economy. Although initial predictions from available Regional Climate Models show substantial spatial differences in projected rainfall changes, there are clearer projections on rising trends in surface air temperatures. There is likely to be rise in temperatures, experienced particularly in northern parts of the country. Although predictions are unclear on any changes in the total volume of rainfall received, some changes in patterns and timings are also very likely and there seems to be reasonable consensus on the fact that the hydrological cycle will be affected. Work done as a part of India's first NATCOM clearly indicates that there will be changes in most river basins in terms of floods, droughts and other river basin parameters. This obviously, has potential for impacting livelihoods and survival strategies, particularly for the vulnerable population.

Rise in sea level up to one metre is expected to inundate 1,700 sq km in Orissa and West Bengal. Such occurrences call for adaptive strategies since these have implications for coastal livelihoods (fishing, agriculture) and infrastructure including the siting, development and protection of ports.

All this implies that climate change is likely to affect India's economic growth through its impacts on resources, especially water resources, forests and coastal lands. Agricultural output and infrastructure are likely to be adversely affected while livelihoods for populations situated in vulnerable areas would be threatened. Resources will need to be invested in combating additional sources of threats arising from specific disaster-related impacts such as those on health and drinking water.

Some Economic Imperatives and Future Concerns

At the regional level, India has been at the forefront among South Asian economies in expressing its concerns about climate change. The National Council on Climate Change has been set up to advice on strategy for development that takes into account the concerns relating to global warming and sustainable development. Climate change concerns and the role of clean development strategies are meant to be an intrinsic part of the Eleventh Plan. Among other initiatives, the "Green India" afforestation programme proposes to convert six million hectares of degraded forest land into green areas. A long-term strategy to deal with glacial melting of the Himalayas is also planned. The National Action Plan on Climate Change discusses the principles and the approach to be followed in meeting the challenges posed by climate change. It dwells on eight national missions in this context. It recognises the role of various social and economic mechanisms such as development of new regulatory structures, appropriate entitlements and pricing.

India's first NATCOM 2004, was a clear demonstration of its willingness to engage with issues of climate change, and a large number of research studies were taken up to assess vulnerability and adaptation to climate change. The MoEF has initiated a second phase of the national communication which will further our understanding of the issues and challenges faced within the economy.

The SRES, IPCC 2007 brings into focus the inter-relationship between anthropogenic forces and climate change, and how these are likely to evolve under various sets of assumptions. It serves to bring into focus the importance of demographic, economic, technological and social factors in determining the alternative future scenarios that economies may face due to global warming. The AI scenario which speaks of a future world with very rapid economic growth alongside the rapid introduction of new and more efficient technologies, which build on convergence among regions and substantial reductions in regional differences in per capita incomes, is a most likely scenario for India in the next couple of decades.

Recognition of the relationship between development and energy consumption strengthens India's right to claims on the ecological space. Energy security is an imperative that affects India significantly and will hold the key to her future economic growth. Energy is essential for poverty alleviation and further development. Arguments are being put forward that the current averages of energy consumption mask the fact that energy consumption levels are in fact skewed, being much higher among the affluent, while those of the poor are much less and need improvement in order to ensure a certain acceptable quality of life. We need to grow, and discussions of equity at the global level must take note of our developmental imperatives such as poverty eradication and attainment of a minimum quality of life. At the same time, on the domestic front, an inclusive development framework must encourage all possible mechanisms for tackling climate change. In doing so, the contentious issue is whether it is possible to go beyond adaptation. At the present stage of development of the Indian economy, this is a crucial question to which there are no easy answers. However, the question has to be asked and attempts to answer it need to be encouraged through further research studies.

The focus should be on what economic growth strategy can best incorporate climate concerns. The two key strategies are of mitigation and adaptation. Mitigation comprises policy and technological interventions to reduce emissions of GHGs. An open economy framework which combines concerns for sustainable development is most conducive for successfully adopting mitigation strategies which would include conservation, reforestation and energy efficiency. Economic instruments can play an important role in promoting adoption of such strategies. GHG mitigation has dominated climate policy internationally, particularly in addressing the issue of decarbonisation of energy systems.

Adaptation on the other hand, implies measures and policy responses to reduce the potential impacts caused by climate change. Adaptation strategies are strategies that may be alternatively called coping strategies. Such anticipatory adaptive measures need to be planned for and not left to the market since market response may be slow and inadequate. The market may fail to provide the right signals for the adoption of adaptive measures and the effects may be compounded if institutional mechanisms are not in place to promote them. Increasingly, impact assessment research for climate change has become oriented towards vulnerability assessment, with a corresponding stress on adaptation strategies in the developing world.

To quote from the Eleventh Plan document, "We have to evolve a positive stance toward adaptation and also constructively with the international community to reach a consensus on mitigation based on fair principles of burden sharing." Further, it states that, "The government's effort will be towards creating an enabling environment for encouraging energy and carbon efficiency in pubic and private sectors and towards internalising climate change concerns in development planning."

Adaptive capacity is determined and can be influenced through various channels including wealth, technology, infrastructure, knowledge and skills and institutions. The role of economic policy is therefore two-fold from a human security point of view—to limit the impacts and also to strengthen adaptive capacity through any of the above channels. Thus, economic instruments can be used for promoting technology adoption, for instance, through carbon taxes, subsidies, etc., as well as promote coping mechanisms through appropriate infrastructure investment.

It is also argued that our ecological footprint is small, we are also improving our energy efficiency and in terms of energy intensity, we are doing relatively well. Historically, industrialised and developed countries have been responsible for 63 per cent of cumulative net carbon emissions from fossil fuel combustion and land use changes since the 1900s with the US being the largest emitter (particularly in the electric power sector). This is often countered by the developed countries with the argument that China and India have contributed significantly to the growth in CO_2 emissions in the Asia-Pacific region with 7 per cent and 2 per cent contribution, respectively to atmospheric carbon content since 1900. Combined emissions from China and India are projected to grow by more than 4 per cent annually between 2010 and 2025. However, it is interesting to look at cumulative emissions since these are a much better indicator of a country's relative contribution to atmospheric concentrations of greenhouse gases from the perspective of economic development. Researchers have shown that even in 2025, with increased emissions from China and India, the US will be responsible for twice as much cumulative emissions since 1980, as China and India combined. Thus, from the equity perspective, developing countries' emissions cannot be compared in terms of responsibility for global warming. However, internally we also need to reduce the gap between the rich and the poor. Although it is not a function of international relations, we must accept that some sections of our rich may be having lifestyles and emissions at par with that of the advanced developed countries. Over 70 per cent of our people are agriculture-based. They in any case contribute the least to emissions. What we need to do is also to look into our own emissions by the rich and have policy measures to keep them at reasonable levels.

Comparisons with China too reveal some interesting facts. Total CO_2 emissions from fuel combustion in India were less than a third of that from China in 2002. China alone accounted for more than half of the total emissions from fuel combustion in Asia. Emissions from OECD countries in North America were more than double of that from Asia excluding China. In fact, improvements in fuel-mix and energy efficiency have enabled India and China to reduce the amount of CO_2 emissions from fuel combustion per unit of GDP over the period 1990-2002 in keeping with the world trends.

Keeping the above in view, the arguments by India holding the developed world primarily responsible for the threats posed by climate change and insistence on them meeting their emission targets is justified. It further strengthens India's calls for facilitating technology transfer and the evolution of mechanisms such as CDM. The Eleventh Plan mentions the use of CDM as an incentive.

At the same time, given the developmental objectives of the Indian economy, particularly in emphasising an inclusive and balanced regional growth strategy, economic policies that encourage the adoption of climate-friendly pathways need to be prioritised. It is human security that is central to this position on climate change. Understanding the links between impact of the climate on ecosystems and human adaptation to these, is a challenging task which require several inputs. These include an understanding of the impacts of climate change on the ecosystems under consideration, and an understanding of the political and economic governance systems that can influence these changes and respond to them.

Economic Aspects of Key Climate Impacts

In planning for resources in the years to come, planning for socio-economic scenarios with and without climate change impacts needs to be done for different time scales, the short, medium and long-term. Climate change impacts such as increased cyclonic activity and changes in temperature and precipitation patterns will have economic implications for these future scenarios in terms of migration and displacement of population, water stress and threats to agricultural production and infrastructure destruction. Sectors that are likely to be affected and which may call for regional re-distribution of resources include public sector utilities, mining and metals, insurance and construction activities. Agro industries, urban and rural infrastructure and tourism are also likely to be affected. These have major implications for India's policy on balanced regional development since certain regions are geographically more vulnerable to climate change impacts than others within the economy. As we incorporate climate change concerns into resource allocation decisions, it implies changes in patterns of production and supply and demand side factors. These in turn will mean tensions as the economy grapples with changes in industrial structure, employment and regional balance in the short to medium term.

Agricultural output and livelihoods of agriculture-dependent populations is likely to be among the most affected by climate change as per available evidence, due to changes in precipitation, temperature and the timing of the monsoon. The occurrence of extreme climatic events implies that additionally disaster-related economic impacts would affect livelihoods in this sector. Therefore, alternative strategies for the future and integration of climate concerns need to be developed which incorporate economic factors such as economic growth, population growth, occupational structure, urbanisation, rural-urban disparity, demand for food and water and trade parameters (for instance, export/import of agricultural or fishing produce). Preliminary research indicates that the regionally differentiation within the country will have impact on productivity and crop pattern due to changes in water availability (including precipitation), solar radiation and prevalence of pests. This could impact both overall agricultural output and throw up distributional equity issues. The Eleventh Plan recognises that there is likely to be an increase in mean temperatures with adverse effects on food grain production with the present level of technology. UNEP estimates indicate that there would be net gains in potential rainfed cereal land for China and Central Asia, and net losses for South and Southeast Asia including India, due to climatic factors. According to the Stern report findings, melting glaciers would increase wet season flooding and reduce dry season water supplies to millions of people in the Indian subcontinent. On the basis of physical and biological principles, the impacts would vary, for instance, in cooler regions low levels of warming could improve conditions for crop growth extending the growing season and opening up new areas for production. However, increasingly negative impacts could emerge from further warming even in these regions. Changes in precipitation and monsoon reliability could have major implications for agricultural production, leading to impacts on the agro industry as well.

Climate change impact studies on agriculture have mostly relied on cross-section data in estimating changes in farm revenue that occurs due to variations in temperature and precipitation. Studies also indicate that climate sensitivity of agriculture is increasing over time. The imperatives for adopting strategies for moving to a low carbon growth path may stem from changes in the global economic order, such as international trade-related regulations, or from domestic drivers such as adversely impacted sectors (which could also affect external trade) like agriculture. Geophysical vulnerability, such as that faced in coastal areas will also be a third factor, making it imperative for adopting climate change concerns in economic planning. The key lies in providing the resources and knowledge to prepare for tackling these drivers. Each of these has its own potential implications for disrupting peace and creating conflicts at different levels. Adapting to climate change and the costs it can impose on the economy is a challenging task. In the context of future impacts through trade for India, it requires the following: a) building a clear understanding of the sectors through which trade-related issues may arise, b) identifying the mechanisms and economic governance issues that can influence change to more climate-friendly practices, and c) an indication of the likely costs the economy may face in the short to medium term with and without a movement towards adoption of climate-friendly practices by the identified sectors. The rationale for this is that environmental policies of major export markets can restrict developing country market access, given that it is feared that external environmental policies could be used for protectionist purposes and that compliance with such policies and measures may entail additional costs.

Rising temperatures and altered water availability would have implications for health status and provision of health care facilities, as it can potentially increase the incidence of infectious diseases such as diarrhoea (water-borne) and malaria (vector-borne). Rising water stress has major implications for urban and rural habitations, leading to another potential source of internal conflict. As states compete for resources to achieve their objectives of economic development, climate change impacts will add an additional source of stress leading to heightened interstate conflicts on sharing of resources—both fiscal and physical—such as water (for instance, through river basin disputes). As policy responsibilities will need to be shared between national, state and local actors, resource allocation issues will become critical.

Adaptation strategies which are relevant in this context include expansion of insurance cover to include climate risks such as flooding, farm practices (cropping changes) and infrastructure. Similarly, siting of industries, planning for infrastructure and location of health facilities should allow for risks emanating from future climate change impacts. The economic valuation of damages and costs of policies to address climate concerns is an important input for designing strategies for preparedness for disasters, insurance schemes, planning for health, water and other infrastructure facilities and siting of industries. An economic and financial question that is being increasingly asked is: Can insurance become an important part of the adaptation strategy? Various experiments in India have been undertaken to provide weather insurance, designed for specific crops and regions.

In a globalising world, with sustainable development as a core goal for the Indian economy, strategies for improving energy efficiency are a must. Improvements in energy efficiency are dependent on the availability of innovative technologies that improve energy efficiency. Both R&D and investments in such technology may be encouraged if distortions in domestic energy pricing (subsidies) are removed. The root causes for increasing emissions lie in fossil fuel combustion. Agriculture contributes less to GHGs. Considering sectoral contributions to CO_2 emissions in India, the highest contributions are from thermal electricity, primarily because of its dependence on coal. However, globally, the amount of CO_2 emitted per kilowatt hour (kWh) of electricity has gone up. Again, the adoption of clean coal technology is obviously a policy that pro-growth options have to consider given the large deposits of coal and our requirements for energy in the coming years. International support is required in addressing how best the developed world (G8) can help in the adoption of such technology. The Eleventh Plan emphasises that, "Development and promotion of low carbon and high energy efficient technologies with reasonable costs will be a priority". Among its environmental targets, the Plan sets a target to increase energy efficiency by 20 percentage points by 2016-17.

To sum, there is a close relationship between climate change and socio-economic concerns for development. As the pressure for cutting fossil fuel-based greenhouse gas emissions increases, parallelly the compulsions of meeting the energy needs of a developing nation get compounded. Thus, climate policy and international negotiations on it require a governance framework (laws, rules and regulations) that is underpinned by the rationale that populations in the developing world will need to move up the energy ladder in improving basic standards of living. Matters may get further complicated for the Indian economy due to trade and trade-related barriers, migration and the "common concerns" of sharing available food, water and energy resources with other developing countries in the region. We need economic policy to take note of both: a) limitations to growth and human well-being that may arise from climatic factors and related international pressures, and b) the mechanisms to overcome these through adoption of new strategies (e.g., technology transfer, R&D, increased energy efficiency).

7 Impact on India's Bilateral Relations with Neighbouring Countries

Uttam Kumar Sinha, Sreeradha Datta, Sunil Chauhan and P.K. Gautam

Overview

The climate crisis in South Asia, a predominantly agro-cultural region with a large population dependent on subsistence agriculture, is likely to create profound challenges. With a population of over 1.3 billion, depleting resources and poverty, the region is vulnerable to climate change and with the increase in frequency and intensity of extreme weather conditions, an unimaginable number of people will be exposed and affected by it.

The impact of climate change, in particular, on water resources and food production will be severe. Water and agriculture are closely tied to livelihood and economic activities. This is the core vulnerability. South Asian states will require implementing development strategies in the farming sector at the national level as well as strengthening diplomatic relations for improving information sharing and cooperation at the regional level. One important set of social impacts resulting from climate change in South Asia will be the potential displacement of a vast number of people as a consequence of rise in sea level.

Changes in temperature and precipitation patterns and numerous other factors will not only impact the natural system but also the human systems including health, displacement and settlement. Investment in mitigation as well as ways to adapt to the unavoidable challenges should go simultaneously with addressing the regional impact that in all likelihood will be created by climate change. A preventive regional policy will be of paramount importance since climate change will heavily impact South Asia's natural environment and nearly all sections of society. Climate change, while not directly related to conflict, can exacerbate existing trends, tensions and instability and thus, should be viewed as a 'stress and strain' factor. The critical challenge is that climate change has the propensity to overburden States which are already fragile and conflict-prone. One cannot ignore that such risks are not only of a humanitarian nature but can also include political and security risks that directly affect State interests. Moreover, it is evidently clear that many issues related to the impact of climate change on regional security are interlinked, requiring comprehensive policy responses. For example, the attainment of the developmental goals in the region would be at considerable risk because climate change, if unmitigated, may well wipe out years of development efforts.

The following section will look at the country-specific impact of climate change and how it will reflect its relations with India.

Bangladesh

Bangladesh has an area of 147,570 sq km and a population of over 153.3 million (2005) with a density of 1209 per sq km.[1] Eighty per cent of the land in Bangladesh is less than 12 metres above sea level. Its population is projected to reach 180.1 million in 2015.

Impact of Climate Change

Sea Level Rise: In February 2008, the IPCC warned that warming ocean temperatures and faster glacial melting would substantially raise sea levels. According to a study, the sea level has been rising in Bangladesh since 1993. There has been an increase of around 4 mm in Hiron Point coast, 6 mm in Char Changa and 7.8 mm per year in Cox Bazar.[2] The rise in sea level rise of 0.5 metres over the last 100 years has already eroded 65 per cent of the landmass of 250 sq km of Kutubdia, 227 sq km of Bhola and 180 sq km of Sandwip Island.[3]

---

1. Agrawala, Shardul, Tomoko Ota, Ahsan Uddin Ahmed (2003). "Development And Climate Change In Bangladesh: Focus On Coastal Flooding And The Sundarbans", Joel Smith and Maarten van Aalst, *Organisation for Economic Co-operation and Development.*
2. Ali, Anwar (2000). "Vulnerability Of Bangladesh Coastal Region To Climate Change With Adaptation Options", Dhaka: Bangladesh Space Research and Remote Sensing Organization.
3. Government of The People's Republic of Bangladesh (2007). "Climate Change and Bangladesh", Climate Change Cell Department Of Environment, Dhaka, September.

Floods and Cyclones: Frequent floods have resulted in flooding of 40 per cent of the total land area annually. On a yearly basis, river erosion washes away 1 per cent of arable land. Bangladeshi scientists estimate that up to 20 per cent of the country's land may be lost to flooding by 2030. About 84 per cent of the total Bangladeshi population is in the rural sector, with nearly three-fifths (61 per cent) of the total land area under cultivation—primarily paddy fields—which occupies four-fifths of the area and provides three-quarters of agricultural produce. On an average, during the period 1962-1988, Bangladesh lost about half a million tonnes of rice annually as a result of floods, especially when rice accounts for nearly 30 per cent of the average annual food grain imports. A study carried out under the BDCLIM (Bangladesh Climate) project sought to examine possible changes in flooding in Bangladesh under a given climate scenario. The studies indicate that the average annual runoff in the Brahmaputra basin would decline by 14 per cent by the year 2050 as a result of climate change.

Sundarbans: The Sundarbans, which straddle southwestern Bangladesh, is a key vulnerable area. With a total area of over 10,000 sq. kilometres, the Sundarbans constitute the world's largest contiguous mangrove ecosystem. Approximately 60 per cent of the Sundarbans fall in Bangladesh and more than 8,000 families depend directly, and at least five million indirectly, upon the Sundarbans. But with occurrence of catastrophic tidal waves, salt contamination has increased, thereby adversely affecting rice cultivation and livelihood. Many industries which depend on raw materials from the Sundarbans will be threatened with closure and create large-scale unemployment.

River Erosion: Riverbank erosion displaces more than 100,000 people annually in Bangladesh. Around one million people have been rendered homeless due to river erosion in Bangladesh. According to a report, the Brahmaputra-Jamuna, a major river system in Bangladesh, has widened to 11.8 km from 8.3 km in the early 1970s, thereby eroding about 87,790 hectares of land. Major rivers such as the Brahmaputra, the Padma and the Meghna have submerged several thousand hectares of land. The number of families and villages which lose their homes permanently to rivers every year are perhaps one of the highest in Bangladesh. It has been reported that many of the slum dwellers in the metropolitan areas are the victims of riverbank erosion. In the decade of 1982-1992, over 106,000 hectares of land has been eroded in the three major rivers of Bangladesh (the Ganges, the Brahmaputra and the Meghna) against an accretion of only 19,000 hectares and about 350,000 people were displaced in that decade. According to a recent study by the Centre for Environmental and Geographic Information Service (CEGIS), erosion along the river banks will result in 29,000 people losing their homes within the next year (2009-10).[4]

Impact on India

India and Bangladesh share the longest land boundary in South Asia. The question of border interlinks with migration and insurgency has a direct impact on bilateral relations. The issue of illegal Bangladeshi migrants is a prickly one. Despite the best of intentions, prejudices and preconceived mindsets have often coloured the estimates of such migrants.

Migration and Security: At the time of the partition, there was a large exodus of Hindus from East Pakistan and the subsequent unfavourable political conditions under Pakistan further led to the subsequent migration with 10 million Bengalis taking shelter in India.[5] About 5.3 million people from East Pakistan crossed over to India between 1946 and 1970.[6] Another large exodus occurred in 1971 just before the Bangladesh liberation war. The real problem is the large-scale influx of economic migrants from Bangladesh into India. The continuous flow of illegal migrants from Bangladesh has resulted in significant demographic alternations, social unrest, political instability, acts of violence and poses a serious security threat to India. Geographic proximity, porous borders, poor border management, lack of employment opportunities in their hometowns and various forms of kinship ties enabled, facilitated and promoted larger scale illegal migration from Bangladesh.

During the 1990s, the cause of migration was economic. Both Hindus and Muslims migrated but Hindus migrated most especially between 1991-1996, during the rule of the coalition government led by the Bangladesh

4. "Bangladesh: River refugee numbers continue to swell", IRIN (UN Office for the Coordination of Humanitarian Affairs), May 29, 2009, *http://irinnews.org/Report.aspx?ReportId=79569*
5. De Amalendu (1988). *Prasanga Anupraprabesh* [essays on Indo-Bangladesh Demographic scenario and influx from Bangladesh] Barna Parichai, Kolkata, p.7, out of the 10 million people, of which 85.5 per cent were Hindus while the rest comprised of Buddhist, Christians and other tribal minorities.
6. Ibid.: p.6.

Nationalist Party (BNP). Even as early as 1971, in a study conducted by Indian Statistical Institute, many experts cited economic and job opportunities in West Bengal as a cause of migration, although a majority believed that insecurity/communal tension was another important cause of illegal migration from Bangladesh to West Bengal in 1971.

While the number of Bangladeshi migrants in India is not certain and varies according to different estimates that range from 5 to 20 million, most of them would come into the category of economic migrants. Unfortunately, there have been no studies undertaken to examine the number of Bangladeshis migrating to India on account of climate change. Despite a number of studies, the issue of migration is fraught with difficulties. First, there is very little data available, which would substantiate the actual number of migrants, as there is a problem/ difficulty in identification of such individuals. Second, apart from certain restricted sample survey studies, which would run into errors, most of the studies have not differentiated the causes that have given way to migrant movements.

Given the social, political and economic conditions in Bangladesh, it would not be difficult to conclude that migration to India is unlikely to decrease. While the exact number of illegal immigrants from Bangladesh is not known, according to the Director General, Border Security Force, the number of Bangladeshi nationals who disappeared after entering India on valid travel documents would be around 12 lakh.[7]

The common borders also give way to other concerns, rivers with its meandering course is one such issue. The trans-boundary rivers keep changing their course and creating fertile islands, for example, the New Moore Island. The Bangladeshis in the Joint River Commission (JRC) meeting in 2006, raised the issue of losing 900 acres of land due to erosion of the border rivers, Ichhamati and Kalindi in Satkhira. Villagers from both sides have lost land and houses due to erosion over the last three decades or more.

Given the continued erosion, these islands called Charlands, emerge in mid-river, with border forces from both sides staking claims on that

7. "12 lakh Bangladeshis went missing in India between 1972-2005: BSF chief", *The Times of India* at *http://timesofindia.indiatimes.com/India/12L_Bangladeshis_went_missing_in_India_between_1972-2005_BSF_chief/articleshow/3405392.cms downloaded on August 25, 2008.*

part of the land. According to the Bangladeshi Water Resources Ministry, Bangladesh has lost around 30,000 acres of land to India due to erosion of different border rivers since the Mujib-Indira Treaty of 1974. The Belonia sub-division in South Tripura, for instance, has been a repeated flashpoint in recent times because the Muhuri river keeps changing its course, creating vast islands, which both the countries claim and which often leads to border skirmishes.

Given the lack of effective border management, cross-border movement of migrants remains a contentious issue. With Bangladesh in a perpetual state of denial over this issue, the Indian solution has been to simply erect fences on the border. To what extent the cross-border movement of people has been resolved is a matter of debate. Given the northeastern states' uneasiness over 'foreigners', the issue of migration from Bangladesh will continue to evoke considerable concern. The 'sons of soil' sentiments in different parts of India will further exacerbate the issue of migrants. The problem is two-fold: migrants' corner the limited economic opportunities causing tension between the local and migrant population and secondly, the migrants become a security risk.

Majority of migration to India emanates from economic necessity and often environmental causes are intertwined with those factors. As estimated by the Bangladeshis, there is likelihood of more than 20 million people becoming environmental migrants by 2050 owing to climate change. The impact of climate change in Bangladesh in the coming years will result in a greater number of migrants entering India and consequently a hardening of perceptions in India. But in recent times, the growing trend in Assam and Tripura and more recently in Maharashtra of evicting those seen as non-indigenous, has impacted not only the population that has migrated 'within' India but also those that have crossed over from the South Asian region as a whole. More importantly, several of the terror attacks that have taken place in different parts of India in the last decade have been carried out by operatives belonging to the neighbourhood countries, Pakistan and Bangladesh.

Myanmar

While there is no direct impact of climate change on the flow of migrants from Myanmar, migration *per se* would still be critical to India's security interest. The steady exodus of political dissidents and ethnic minorities from Myanmar into neighbouring Thailand, China, Bangladesh and India, has been quite marked ever since the State Law and Order Restoration Council (SLORC) took over control of the governance of Myanmar from the Burma Socialist Programme Party (BSPP) in 1988. Close historical and cultural ties across the 1760 kilometre Indo-Myanmar border facilitated the flight of thousands of Myanmarese nationals into Manipur and Mizoram. The first phase of the exodus in 1988 consisted mostly of student leaders and political dissidents. However, over the years, the two northeastern states have given refuge to politically persecuted members of the ethnic groups, mainly the Chins and Arakanese community, people from the Chin State and the Rakhine State.

According to some estimates, for the past 20 years, Mizoram, with a population of approximately nine lakh, has been hosting the Myanmarese population, estimated between 70,000 and 100,000. Resentment against the 'outsiders' is gaining momentum in Mizoram and the Young Mizo Association has spearheaded the campaign to evict the Chins from Mizoram. According to the World Refugee Survey 2008, there are many Muslim Rohingyas in the Andaman Islands, often arrested for arms smuggling and other charges.

With an increase in natural calamities and the impact of climate change, the number of migrants from Myanmar into India would increase several-fold. Apart from demographic tension that such migration will invariably lead to in the northeastern states, the spread of HIV/AIDS and increased drug trafficking from the region will add to the complexities.

Pakistan

Probably nowhere in the region is the 'climate change-water resources-food production connect' as deep and worrying as in Pakistan. It faces one of the severest water shortages in the world with a per capita availability per annum of 1100 cu. metres. This figure is alarming given that it is below the internationally recommended level of 1500 cu. metres and precariously close to the critical 1000 cu. metre level. In 1951, the per capita availability of water per annum was 5300 cu.metre. That is indeed a quantum fall. Many factors contribute to this critical condition. Declining flows in the Indus river system, the elusive and contentious inter-provincial water accord due to mutual suspicions among provinces and an unsustainable population growth rate of 2 per cent, all combined to not bode well for Pakistan's water situation.

Disagreements on construction of new reservoirs, declining groundwater potential and growing number of disputes with India after a relatively uneventful 44-year period of water sharing under the Indus Water Treaty (IWT) will further complicate matters. In sum, the water situation in Pakistan (a country whose landscape is largely arid to semi-arid) is truly disastrous in spite of the Indus, its tributaries and a treaty with generous concessions that has been implemented by India. Pakistani farmers may be forced to change to high-yielding early-maturating crops, modify their sowing patterns and employ micro-irrigation in the coming years to mitigate shortage—all of which will entail higher costs.

Pakistan will continue to build pressure on the Indus waters. Its objections to Indian projects and a general unwillingness to engage India constructively are partly to force India to amend the IWT to accommodate the emerging patterns of water-use in Pakistan, such as water-sharing during periods of shortage, a situation not envisaged in the treaty.

The treaty will be tested under the bilateral strains and domestic compulsions. Questions will be raised on whether the treaty, under changed hydrological circumstances, has outlived its utility. Does it need to be renegotiated, possibly even negotiated afresh? Will the IWT have the steam to galvanise an IWT-2, structured on new arrangements (Article VII of the treaty is about future cooperation), thus taking into consideration the domestic needs on either side?

Future initiatives on the IWT will require a convergence of political will on both sides. In the meantime, the *status quo* will prevail, an acknowledgment and acceptance, however grudgingly, of working within the framework which neither party would like to give up. However, that said and given the nature of subcontinental politics, there will be an increasing use of water as a 'tool and a bargaining instrument' in the larger politico-strategic objective. Pakistan will shift the focus on Kishenganga, since the neutral expert Raymaot Laftee gave a clean chit to the Baglihar project in 2007. Pakistan's expression will be on similar lines, of what can be termed 'lower riparian angst', vulnerable lower riparian using international platforms to air its grievances. On the Kishenganga, Pakistan's concern is that India's project plans and diversion will reduce the flow of the Neelam (Kishenganga) by about 27 per cent and the direct impact will be on its agricultural usages. It will, like it did on the Baglihar, stonewall and raise objections to the project and seek international attention. As disputes over the treaty magnify, particularly the distributive issues of river waters—getting more of what is in dispute—the politicisation will be high and a search for a resolution on the Indus water issue will form the core of the Indo-Pak relations.

Island States

Sri Lanka

Sri Lanka is a typical island nation with an extensive coastline, about 1,585 kilometres and a large coastal zone. The coastal zone consists of 24 per cent of the land area, accommodates 32 per cent of its population and accounts for 65 per cent of its industrial output.[8] With a large and extremely fragile coastal zone, Sri Lanka is highly vulnerable to any significant rise in sea level due to climate change and associated coastal erosion. The country has been experiencing coastal erosion at the rate of 0.30 to 0.35 mm per year for 45 to 55 per cent of its coastline already[9] and increased acceleration due to climate change will lead to increased loss of land, and affected communities and economic activities. Country estimates of land loss and inundation resulting from rising sea level scenarios ranging from 0.30 to 1.0 mm would lead to land loss between 6.0-11.5 sq. kilometres and inundation of area up to 41.0-91.25 sq. kilometres.[10] The other effects of climate change would include increased flooding, storm damage and breakdown/slow natural alterations of sea defence structures and port infrastructures.

Impact on India: Sri Lanka is a democratic State but challenged by separatist forces, and relations with India are largely influenced by the ongoing civil war and consequential cross-border ramifications, besides illegal poaching and drug/arms running.

8. Ministry of Home Affairs (2002). First National Communication of Sri Lanka to the UNFCCC. p.64.
9. Ibid.: p.66.
10. Ibid.: p.66-67.

There are no immediate climate change-related threats to the India-Sri Lanka security calculus but an adversely impacted coastal zone could affect the stability in the Gulf of Mannar, which could become a hotspot of increased cross-border conflicts through heightened incidents over illegal fishing and plundering of maritime resources. With the Sethusamudram Canal project in the sensitive Gulf region, increased shipping activity would provide easy targets for rival groups in conflict over resources.

Other potential sources of conflict could be future disputes over claims to maritime zones and resources. Sri Lanka has approximately 489,000 sq kilometres of maritime waters and both countries have resolved all claims over maritime waters through two bilateral agreements (1974 and 1976), and one trilateral agreement along with the Maldives (1976). The rise in sea levels could shift the demarcation of baseline points (low water marks), significantly in the case of Sri Lanka, being an island nation and already prone to high sea erosion. Any shift in baseline points would alter the boundaries significantly, potentially changing fishing, mineral and marine rights along the maritime borders and becoming a source of serious contention and conflict.

Both India and Sri Lanka are developing countries and on the issues of climate change, both countries need to coordinate their negotiation positions as they have similar development interests, besides great scientific cooperation in research and development as well as in practical implementation of mitigation and adaptation strategies.

The Maldives

The Maldives, a fragile, low-lying small island ecosystem, is very vulnerable to climate change, especially the predicted sea level rise. Ninety per cent of the Maldivian territory of over 90,000 sq kilometres is under water. Only 202 islands are inhabited by Maldivians. Another 87 islands have been developed as tourist resorts while the rest of the 1192 coral islands are uninhabited. It has been estimated that a two-metre rise in the sea level will submerge virtually all coral islands scattered some 850 kilometres across the equator, forcing its 300,000 citizens to evacuate. Although the Maldives contributes minimally to the global greenhouse gas emissions, it is among the most susceptible to impacts of the changes in climate. The sea level rise projected in IPCC AR4 threatens the very existence of life and livelihood in the Maldives. Many islands already suffer inundation and shoreline erosion because of its low elevation. Inundation often leads to freshwater shortages and outbreak of diseases.[11] It is expected that even a one metre rise in sea level would cause the loss of the entire land area of the Maldives.

Impact on India: Given the friendly bilateral relations between India and the Maldives, in course of the steady rise in sea level, India in all likelihood would have to absorb many of the internally displaced Maldivians. The Maldives is of great strategic significance to India. Located in the Indian Ocean and close to major sea lines of communications, it has unique marine resources. With the US stronghold in Diego Garcia, a favourable Maldives to Indian military presence in its waters could provide a counter-balance to the large number of navies that are operating in the Indian Ocean for safeguarding their maritime sea lines. With China already making overtures for infrastructure development as well as environment preservation projects in the Maldives, it is emergent on India to undertake radical shifts in its policy towards the Maldives.

Climate change offers a unique opportunity to India for a cooperative engagement policy with the Maldives and for enhancing its foothold in the Indian Ocean region:

Alternate Sources of Energy

In order to reduce the emission of greenhouse gases, the Maldives has started pilot projects on alternate sources of energy. Solar power has been used to power telecommunication sets, navigational aids and government office buildings and mosques in the islands. The main constraint to the widespread use of solar energy is the lack of technical backup and the high installation costs. Solar energy is one of the national missions of India as enumerated in the recently released National Plan. Indian efforts should include requirements and participation of the Maldives with assured sharing in development and full commitment to implementation. Further, supplementing conventional energy supply by alternative energy sources wherever viable, has been included in objective for the energy sector and the strategy in the National Development Plan of the Maldives, and India,

11. Ministry of Home Affairs, Housing and Environment (2001). *First National Communication of Maldives to the UNFCCC*. Republic of Maldives. p.5.

with its developed technological base could provide the Maldives with support.

Security from Eventuality of Island Sink

The Maldives has a population of 300,000 with a projected growth rate of about 2 per cent. The average population is very young, with a majority being below 30 years of age. The dominant concern among the young is of the eventuality of evacuation to safer zones in the event of a one metre rise in sea level. India, the most proximate of nations—both physically as well as historically and culturally—is well placed to offer sanctuary to the climate migrants. The complex challenge needs to be studied deeply and methods of evacuation, localisation of migrants and their eventual integration with our communities need to be worked out.

Infrastructure Protection and Combating Erosion of Islands

The Maldives is looking for solutions to combat coastal erosion and lacks technical and engineering expertise to prevent further erosions. There are attempts at raising one atoll by three feet in a joint process, with India as one of the participants.

Landlocked States

Nepal-Bhutan

According to the IPCC AR4, climate change will increase risks in Nepal and Bhutan, particularly as the Himalayas will experience rapid glacier melt with a rate of recession higher than anywhere in the world. While climate change will have an interlinked impact, the two most vulnerable areas in these two landlocked countries will be agriculture and water resources. In Nepal, agriculture would be adversely affected by the loss of the fertile top soil due to soil erosion, landslides and floods. Soil loss could be a major cause of decline in agricultural production. Water crisis would be high particularly in the dry season. The demand for water in Nepal is estimated to increase seven-fold in the next 25 years. Some of the adaptation efforts would require close cooperation with India not only in terms of building dams, reservoirs and hydroelectricity projects but also for basin-scale adaptive measures on the rivers Karnali, Narayani and

Kosi. This would include development of disaster mitigation measures, for example, the siphoning of lake water, establishment of a flood alarm system and construction of an outlet channel.

The impact of climate change in Bhutan would be two-fold. An increase in temperature by 2°C would shift the cultivating zone further into higher elevation, resulting in adaptation for both crops and livestock. Low-income rural populations that depend on traditional agricultural systems or on marginal lands would be particularly vulnerable to climate change. Water resources, a critical area, would be sensitive not only to the changes in temperature and precipitation but also to changes in tropical cyclones. Reduction in the average flow of snow-fed rivers, combined with an increase in peak flows and sediment yield, would have major impacts on hydropower generation, urban water supply and agriculture. Glacial lake outbursts would include perturbation in the quantity of river water used for hydropower generation, destruction of settlements, infrastructure and agricultural lands, and loss of biodiversity and even human lives downstream.

China

The impact of India's bilateral relations with China can be analysed from five levels:

Convergence and Divergence in Climate Change

Both China and India in the recent past have released policy documents on climate change. One common understanding is that they both oppose securitising the issue and its inclusion in the United Nations Security Council (UNSC). Both countries are also part of informal agreements on climate change such as AP6 (US, Canada, Japan, Republic of Korea, Australia, China and India). They are also under the G8 Heiligendamm process as part of the O5 or Outreach 05 (with Brazil, Mexico and South Africa) and also members of the Major Emitters Meeting (MEM 16), i.e., G8, plus O5, plus Indonesia, South Korea and Australia.

Negotiating Position

While both the countries are exempted from the binding carbon emission reduction under the Kyoto Protocol, clubbing the two in the climate change debate with the developed world might have strategic relevance at the tactical level. India would like to delink itself as its per capita carbon dioxide emission is far less than China. The 2004 figures put the per capita carbon dioxide emission (in metric tonnes) at 3.8 for China and 1.2 for India. India emits just 4.6 per cent of the total carbon emission and China, in contrast produces 17.3 per cent. Though India has stated that its per capita greenhouse gas emissions will at no point exceed that of developed countries, the question of total population and total emissions may be questioned by the international community.

Qinghai-Tibet Plateau

The rivers emanating from Tibet and entering India will be a key driver of bilateral relations. China has grand plans to develop this region which is bound to conflict with the Tibetan model of development based on environmental safeguards and sustainability, and a participatory 'need-based' development. The Chinese developmental path will lead to a certain damaging impact on the Himalayan and grassland ecosystem. China thinks and executes gigantic projects, often subduing nature. The physical impact of abnormal glacier melt leading to incidents such as flash floods in the Tsangpo and Parechu in 2000 and 2004 will be issues which would need a high level of cooperation. Extreme floods and their increased frequency would raise the level of distrust and tension.

Water Sharing Issues

Perhaps the most challenging issues will be over-sharing of water from rivers flowing into India from Tibet. Geography has an important role to play; unlike India, which is both an upstream and downstream country, China due to its geographical location, is an upstream nation. Its policy orientation is driven by its upper riparian status. For its economic and population growth, augmenting its water resources is a top priority. Three water diversion channels to the north are planned from Yangtze river to the Yellow river. Upon completion by 2050, a total of 44.8 billion cubic metres of water (equal to the water flowing down the great Yellow river each year) will be transferred by this man-made river. Like India, China is also a victim to climate change. One scenario predicts decrease in precipitation in north-central and northeast China from 2020 onwards.[12] An official Chinese study shows a 37 per cent drop in precipitation in the Huai, Liao and Hai rivers in the near future. The report also mentions glacier melt-overflow, leading to floods and then drying up for the Yangtze and Yellow rivers.[13]

Diversion of the Yarlung Tsangpo branch of the Brahmaputra around the famous U-bend before the river enters India is worrying. Though assurances have been given that it is not planned, due to climate-related water stresses, it may force China to ignore lower riparian rights. Similarly, glacier lake outflow floods, both in Brahmaputra or Sutlej, may increase. Even if they are not by design but by nature, the issues would require a high degree of data-sharing, trust and cooperation.

WTO

Climate change and international trade are inter-related. Carbon intensity of products is surely going to be on the trade agenda. The global market for environmental goods and services is estimated to be worth more than $550 billion every year.[14] Environmental goods and services and barriers to trade due to climate-related technologies are the areas where more research and clarity is needed. There are bound to be areas where India and China may differ.

---

12. Harrison, Stephan (2005). "Climate Change, Future Conflict and the Role of Climate Scientist", *Royal United Services Institute Journal*, December, pp.18-23. Author quotes data from *Climatic Change* (1999).

13. Economy, Elizabeth (2007). "China's Coming Environmental Crash", *Foreign Affairs*: 38-57, Sept.-Oct.

14. Vasudeva, P.K. (2008). "WTO Accord", *Yojana*: 18-20, June.

8 Impact on Warfighting Capability of the Indian Military

Sunil Chauhan, P.K. Gautam and Ajey Lele

Introduction

In one understanding, securitisation and militaristic thinking on climate change is counter-productive. Even conceptualisation of climate security, as is being proposed and propounded by developed countries, is another tool of the geopolitical power game.

In military matters, it is important for militaries to study the developments in other militaries from open sources. There is some deal of confusion when impact on warfighting capability is compared with security scenarios. For instance, the German Military Staff has included climate change issues in its long-term security scenarios until 2040. The Strategic Future Analysis group defines trends in issues such as demography, development of resources, science and technology, cultural, social, economic and political. The results obtained from trend analysis are then translated into security scenarios and description of fictitious events beyond 15 years. Such studies by the military project scenarios. But how exactly they impinge on warfighting capability is not spelt out.[1] A recent and widely quoted study in the US done by the Center for Naval Analyses (CNA) spells out in general terms issues of energy efficiency and impacts of extreme weather and such events on US military installations over the next 30 to 40 years.[2]

---

1. Jopp, Hienz-Dieter and Roland Kaestner (2008). "Climate Change and Security in Twenty-first Century", in Hans Gunter *et al.*, *Globalization and Environmental Challenges: Reconceptualizing Security in the 21st Century*. New York: Springer. Chapter 72.
2. CNA Corporation (2007). *National Security and the Threat of Climate Change*.

This chapter will analyse the impact of climate change on the capability of the Indian military from the perspective of land, sea and air. It is an attempt to project and visualise how conditions on the ground may change and how to ensure military effectiveness with changing climate pattern. It is intended that the military is made aware of the factors and takes suitable measures on a no-regret options path.

LAND

Glacial Meltdown

Himalayas: Glaciers and permafrost in the fragile Himalayas and Tibetan plateau, already under strain from military deployment, are melting. The meltdown may become very rapid, which has its own military implications such as disruption of communications. Flash floods in the shared rivers may further fuel mutual suspicion between not-so-friendly upstream and downstream countries. We can recall the flash floods of Parechu river of 2000 and 2004 in Himachal Pradesh.

Siachen Glacier: The Siachen glacier is a hotspot in the continuing Indo-Pak military standoff. Demilitarisation is not an option and a ceasefire has been in place since November 2003. While the negotiations would continue, it is important to visualise the impact of the sudden and accelerated meltdown of glaciers in the next 10-15 years. Once the glacier/snow caps melt, only the rocks would remain. Rapid melting would bring in flash floods and a series of mini-disasters downstream. It is unlikely that the existing defence works and elaborate communication infrastructure (on ice) would last after melting. Artillery gun platforms that have become ice pillars would melt rapidly, making redeployment a necessary but demanding task. Unexpected melting would make movement of troops extremely dangerous, if not impossible. Level and firm dropping zones and helipads may crumble with the rapid snowmelt. The army, in any case, is adapting to any changed physical conditions with innovation and adjustments.

Mobility across Snow-Bound Passes

With likely reduction in snowfall due to climate change, mobility across snow-bound passes could improve considerably. Reduced snowfall in a way came to the aid of the Indian military during 'Operation Vijay' at Kargil in 1999. Because of reduced snow, the Zojila axis, which remains inaccessible, could be opened up in early May 1999, well ahead of the annual 'road open cycle' in June. This allowed induction of additional forces by road from Srinagar via Zojila. In a normal year, with the same snowfall pattern, it is unlikely that rapid induction would have been possible. It also came as a surprise to Pakistan and upset their plan to cut off the axis.

However, climate change-induced glacier melt would lead to flash floods and severing of road communication. This would counter the afforded better mobility with reduced snow over the passes. The possible varied outcomes would complicate military planning. With the change (reduction) in snowfall and the threat of melting of the permafrost in Tibet, new environmental/topographical factors need to be considered by strategic planners. The railway track laid from Golmud to Lhasa in Tibet may also be affected adversely. This may upset Chinese logistic plans to speed up build-up in one summer campaigning season. Purely from a theoretical point of view, permafrost meltdown reduces the logistic tonnages of the Chinese and reduces its offensive capability in Tibet.

Spread of Vector-Borne and Other Diseases

With climate change, vector-borne diseases have spread. New strains of germs and viruses may evolve as a response to man's destruction and neglect of nature. In order to be prepared for these emerging threats, a greater cooperation and integration between specialists in animal and human diseases is essential.

Impact at Tactical, Operational and Strategic Levels

Strategic Level

The concept of campaigning season may need a reappraisal due to projected changes in climate and weather.

Operational Level

- More heat may demand higher scales of water in hot weather warfare conditions in operation logistics. Similarly, wear and tear on equipment may be more pronounced due to further increase in temperatures. These need to be studied.

- Glacial lake outburst floods (GLOF) and restrictions on mobility in mountains due to flash floods may require campaign planners to have alternative means of mobility or change of plans.
- China's National Climate Change Programme of June 2007 mentions that the thickness of frozen earth in the Qinghai-Tibet plateau reduced by a maximum of 4-5 metres in the last 50 years. Glaciers in Western China will reduce by 27.7 per cent by 2050 and permafrost will be altered.

Tactical Level

- Redundancy in Mobility: Surface mobility will be impacted in the mountains due to increase in flash floods. Redundancy in mobility by way of rotary wing, road transport, animal transport and foot mobility will demand training and wherewithal. Stocking levels may need to be increased due to more frequent disruption of communications. Greater reliance on animal transport and the necessity to sustain the legendary capacity of the Indian Army to march will need to be sustained.
- Tactical Deception: Forest die-down may require adjustments to camouflage methods.
- Health Precaution: Heath precautions such as hot weather precautions and prevention from vector-borne diseases will need more attention including anti-malaria precautions in higher latitudes/altitudes.
- Technical Gunnery: It needs to be confirmed if there would be a need to take measures in changed firing tables standard conditions for artillery guns, rockets and mortars due to increase in surface temperatures. (FT conditions being 15°C).
- Weather Forecast: More sophistication may be required to forecast weather conditions for military operations.
- Fire Precautions: Storage of ammunition in higher ambient temperatures due to global warming (as also heat island effects in built-up areas) needs to be studied.
- Remote Sensing: Urban pollution, forest fires, atmospheric brown cloud, coastal pollution and algal bloom may have an impact on remote sensing.

Logistics

As we confront peak oil and need for alterative fuels, retrofitting with hybrid/fuel cells or hydrogen economy will need to be done in a phased manner. Renewable solar energy will need to be tapped innovatively. Liquid fossil fuel would continue to be the main fuel, for which biofuel from military land, may be one source.

Role of Scientific Studies

Science will play a critical role in forecasting seasonal patterns and monitoring the impact of global warming. Impact like reduction in the logistic tonnages by surface means (road and rail) and affect on military facilities like airfields will need scientific monitoring and study. What will be the stance in Indo-US relations if the base at Diego Garcia submerges? How the US and NATO power projections will be impacted is another area of study.

Naval Operations

Climate change has the potential to alter the balance of the mix of Indian naval missions and strategic interests, and could even create new mission types. The future presents a range of strategic force structure, capability and energy efficiency challenges and opportunities to the Indian Navy, including responsibility in mitigating climate change by minimising the impact of its activities and driving technological change. The Indian Navy will need to adapt to the future security environment and operational conditions, being shaped by climate change.

Increased Relief, Recovery and Humanitarian Commitments

The impacts of the growing risk of disasters and crisis situations because of climate change and the consequent damaging effects on development, especially in developing nations like India, will increase the pressure for naval forces to participate in growing number of humanitarian assistance, disaster relief and evacuation operations across borders. This has already manifested itself in the major disasters of the past, *viz.*, the Tsunami (2004), and Cyclones Sidr (Bangladesh, 2008) and Nargis (Myanmar, 2008). Participation in increased humanitarian missions will require realignment in operating procedures, training principles, ship design and outfitting.

Force Structuring: To carry out relief missions at short notice, the ability to move rapidly and land large volumes of supplies is crucial. The impacts of climate change could also affect the use and availability of amphibious landing sites. The Indian Navy may have to consider acquiring more shallow draught ships capable of landing in disaster areas and heavy-lift helicopters for ship-to-shore transport. The use of air cushion vehicles and hovercrafts could become a more attractive option. Consideration may need to be given to replace a number of brown water ships with multi-role expeditionary capability warships.

Equipment Standards, Hardware Profiling: If climate-induced disaster relief missions become more frequent and/or demanding, there will be a growing need for naval equipment relevant to these activities. Important equipment to be considered, where Indian naval ships do not have capability currently are: construction, sanitation, fuel transportation, pharmaceuticals and medical supplies, power generation equipment, etc. Lessons learnt from Tsunami 2004, disaster-specific or dual-purpose equipment could become an important consideration in ship design and outfitting.

Ship Design: Rapid response and deployments would require ship designs with greater role adaptability for the crew and the hull. Greater sea-keeping ability may become more important in order to cope with higher sea states.

Ship Manning: The Indian Navy would require carrying humanitarian and civil relief workers more frequently, requiring a shift in operating principles and including them in regular training missions too, in non-crisis periods. Greater relief operation might also call for frequent use of reserves and volunteers. (Tsunami 2004 relief operations had naval ships carrying national and international relief workers to disaster areas and working in tandem. The Indian Navy also transhipped and landed few disaster specialists to Myanmar for relief operations during Cyclone Nargis.) Also, there could be the involvement of various government agencies and authorities too.

Training Doctrine: An increase in the frequency of conducting disaster relief operations could drive changes in the training doctrine of the Indian Navy towards increasing emphasis in including humanitarian missions in training programmes and exercises. Hitherto, the training and exercises are skewed largely towards conventional naval warfare tactics, with humanitarian aspects and evolutions covered largely in classroom instructions. Training people to operate equally effectively in humanitarian missions would require integration in full-scale functional exercises.

Logistics and Communication Capabilities: The Indian Navy's capability for logistics and communication could be stretched.

Sea Level Rise

Sea level rise would affect Indian naval operations at all levels of warfare—strategic, operational and tactical—impinging on issues of maritime boundaries, exclusive economic zone (EEZ), port operations, shallow water operations, notably for submarines and naval tactics.

Maritime Zones and EEZ

Maritime Zones Act: In 1976, the Territorial Waters, Continental Shelf, Exclusive Economic Zone and other Maritime Zones Act came into force. This Act dramatically increased India's responsibility from 83,200 sq. kilometres to an area of 2.02 million sq. kilometres of sea, or about two-thirds of the total area of land, along with the potential addition of 1.5 million sq. kilometres of the continental shelf, which will be finalised in the near future. Any rise in sea level would affect the baselines of these zones. Modelling studies on the likelihood of inundation because of sea level rise along the vulnerable zones need to be undertaken and simulation studies carried out, to generate alternate baseline points for various scenarios of sea level rise based on the coast profile.

Delineation of Maritime Boundary Agreements: India has concluded Maritime Boundary Agreements in pursuance of Maritime Zones Act of India, 1976 and UNCLOS III, with five of its neighbours. India has signed 12 maritime agreements with five of its neighbours on opposite coasts — nine bilateral agreements with the Maldives (1976), Sri Lanka (1974 and 1976), Indonesia (1974 and 1977), Thailand (1978 and 1993) and Myanmar (1986); and three trilateral agreements with Sri Lanka and the Maldives (1976), Indonesia and Thailand (1978) and Myanmar and Thailand (1993). As a result, the delimitation of India's maritime boundaries with four of these five States is complete; the only exception, Myanmar, requires a trilateral agreement to determine its tri-junction point with India and Bangladesh. However, this necessitates the prior resolution of the maritime boundary dispute between India and Bangladesh. The only countries with whom maritime boundary agreements have not yet been concluded by India, are Pakistan and Bangladesh. The zone of contention with Pakistan is the Sir Creek area and the delimitation of maritime boundaries. As regards Bangladesh, there are three major maritime issues which have not found any solution till date. These are Farakka Barrage (not a maritime boundary issue but a maritime issue), New Moore Island and the maritime boundary dispute. Any shift in baseline points would alter the boundaries significantly and become a source of contention.

Port Operations: India has a coastline of 7,516 kilometres with about 230 outlying islands, which makes it the 15th longest coastline. Thirteen major and 173 minor and intermediate ports are located along the coast, with a large number of newer ports under construction. Almost 95 per cent of India's external trade is handled through these ports. Rise in sea level and increased storm surges and cyclones would change the hydrography, hydrology and the coast lining of regions. In port areas, these would have severely adverse affects on shipping operations, necessitating frequent dredging or new hydrographic surveys. The Indian Naval Hydrographic Office would be faced with greater demands because of increased weather extremities. Sea level rise and storm surges could also make naval dry docks unusable and difficult to access because of increased wave heights and changed tidal patterns. The Tsunami in 2004 and the consequent increased sea level in Port Blair rendered the civil dry docks unusable, which had to be modified for the present variations in sea level and tidal reach.

Protection of Bases and Coastal Installations: The Indian Navy would also need to carry out vulnerability assessments of its bases and coastal installations against degradation from the extreme weather conditions and rising sea levels. Both INS Kadamba, the newly commissioned forward naval base in Karwar, and INS Angre at Mumbai are particularly at threat from the predicted increased storm surges and cyclones in the west coast. The location of bases and other shore/offshore assets would also need to be re-assessed, depending on weather risks and operational necessities.

Operational Oceanography

Operational oceanography plays a key role in naval warfare, bearing directly on operations and tactics. Ocean data, *viz.*, temperature, salinity and pressure profiles, ocean circulation patterns, especially of the upper layers of the ocean (up to depths of 2000 metres) govern submarine operations, performance of underwater sensors, *viz.*, sonars and underwater telephones, and firing of weapons.

Climatic change at predicted rates could alter the ocean circulation patterns and bathymetry data significantly. Anecdotal evidence shows that notable changes are taking place in the Indian Ocean and the Bay of Bengal, with observed bathymetry data showing significant variation in the last few decades.

At the tactical level, this would imply more predictive data on bathymetry, ocean currents, surface temperature and the evaporation duct in operational areas. Efforts at better predictions may require augmenting of naval ships' meteorological capabilities. Currently, only major warships are fitted with radiosonde devices; these may require to be fitted even in minor war vessels, especially those operating closer to the coast and environmentally threatened/challenged areas, such as island territories, coral banks and coastal areas.

Operations in Extreme Weather Conditions

Increased extreme weather conditions with rougher seas would also change the navy's exercises, as many—especially weapon firing—exercises have sea, state and weather restrictions, thus reducing flexibility and calling for redesign and restructuring of exercises. Extreme weather conditions could also affect maintenance schedules of ships and require recalibration of refit cycles within the lifecycle of ships. Operating under adverse conditions could seriously affect efficiency of the crew, necessitating frequent turnaround and revising manning policies.

Curtailing Emissions

Currently, naval ships and shore assets are either exempt or overlooked from environmental considerations and emission standards. However, in the decades to come, growing political and economic pressure on government agencies to implement energy efficient technologies and methods could see the Indian Navy working towards a radical shift in its operational procedures. Curtailing exercises, increased use of carbon-free fuels such as LPG, ethanol for non-missions critical equipment, reduction in non-operational energy consumption would impinge on actual operations and delivery of combat power when the time comes. This would require re-working of strategy and tactics so that while the methods change, the desired result is still achieved.

Force Structure Decisions: Indian naval operations and exercise schedules are affected by fuel conservation. Reducing fuel as well as all other forms of emissions is also a tactical requirement for stealth operations as these are vulnerable to detection. The early to mid-1990s saw a dramatic revision of naval exercise patterns with the 'much more with much less' principle dictating that ships compress as many exercises in a day to complete the exercise in fewer days, to avoid excessive fuel wastage. This was stretched to the extent of even curtailing 'illumination' of ships at anchorages during Republic Day and Independence Day celebrations. Climate change will add to this operational challenge to adopt technologies that reduce the carbon footprint of ships. Naval R&D efforts on alternative forms of energy for ship consumption, solar, wind and low emissions fuel are presently minimal, which needs to be reversed. Reports indicate that the US is experimenting with designs of new high-speed sail ships, powered both by wind energy as well as solar energy, for expeditionary operations.

The Geopolitics of Naval Operations

Current trends in naval operations are motivated by concerns for energy security, safeguarding maritime interests and cost-effectiveness. Climate change at predicted rates is likely to intensify these trends and dominate future naval operation paradigms.

Changes in Marine Biosystems: Climate change-induced shifting marine resources, especially marine fisheries, in many cases across international boundaries, could call for revision of maritime zones, fishing zones, coastal navigational routes. Ocean acidification is already affecting fish stock migration and there is likely to be increased illegal fishing in Indian waters, especially regional fisheries. Protecting our marine resources would require increased Indian naval and coast guard assets.

The Arctic Ocean Energy Resource: The rapid retreat of the ice in the Arctic Sea has opened up a series of border and mineral disputes between Canada and the US over the legal status of the Northwest Passage, between Greenland and Denmark over independence and between all Arctic nations over mineral access rights. The Russian Arctic is a storehouse of natural resources. Changing climate will spur an increase in the exploitation of energy and mineral resources. Also, with a projected ice-free Arctic environment, countries such as the US, Russia and the UK are already gearing towards increased operations in the Arctic. As an aspiring global power, the Indian Navy cannot afford to ignore the eventuality that it would need to operate in the Arctic waters, which throws up complex challenges. Modelling for prediction of the environment, ship design modifications, new weapons and sensors and training of the crew for the conditions, will be important.

Naval Budgeting

The impact of climate change on the economy could well curtail the naval budget. Naval operations are energy-intensive, and are already running on a tight budget. Any further curtailing would affect operational readiness. The challenge is to innovate, revise and modify training and operations with reduced carbon footprints for the same level of readiness.

Air

Indian military aviation and air warfare are expected to be affected by both climatic and weather changes. The geographical extent of Indian military aviation tasks stretches across a vast region. As per the Ministry of Defence (MoD) Annual Report 2002-03, page 2, paragraph 1.6:

> India's size, strategic location, trade interests and a security environment extends from the Persian Gulf in the west to the Straits of Malacca in the east and Central Asian Republics in the north to near the equator in the south.

The likely climatic changes over this vast region are expected to be of an unstable nature and difficult to pinpoint exactly at this stage. However, generally speaking, the changes are expected to be on similar lines as the rest of the world, like increase in surface temperatures, seal level rise, glacier melting and increase in frequency of cyclones and probably a shift in the seasonal weather cycle. Broadly, climate change is likely to alter the pattern of the climate in general and this may lead to varying weather conditions. The basic elements of weather namely, precipitation, temperature, pressure and winds, clouds and humidity and sunshine, are expected to show variability as compared to the existing pattern.

Impact of Weather Variability

Military aviation involves various tasks undertaken during peacetime and wartime by the Air Force and the aviation wings of the Army and the Navy. It also includes aid from civil aviation in case of necessity. All missions undertaken for the purpose of military requirements always factor in the weather during planning as well as during the execution stage. For India's military aviation, the challenges are enormous from the 'weather' point of view. The flying is undertaken over an area which has varying weather conditions in all seasons. The performance of the aerial platforms as well as that of the munitions delivered from these platforms varies considerably, depending on ambient weather conditions. Presently, military aviators are trained to factor in varying atmospheric conditions to undertake their various tasks. They are constantly fed with weather inputs during 'planning' as well as during the 'execution of task' stages.

Manned and unmanned aircrafts are expected to operate over the length and breadth of the vast region of the continent inclusive of land and ocean areas, both during peacetime as well as wartime. The aerial missions undertaken would be varied, depending on specific requirements and are expected to include various types of aircrafts like fighters, transport aircraft and helicopters. The operating heights for these aircrafts would be different and hence at times they would experience different weather conditions in spite of being operating over the same geographical region. Military aviation would be expected to play a role in a host of missions including aerial combat, air defence, close air support (CAS), reconnaissance, paradropping missions, transport operations, fire support, aerial refuelling, gathering of electronics intelligence, search and rescue, early warning, etc. It is also expected that for out-of-area operations the Air Force would be heavily depended upon. For all these missions, weather support is critical and impacts both planning and execution.

Presently, military aviation planners include weather conditions in mission planning on a regular basis. Weather also plays a key role in weapon delivery, particularly in the case of standoff weapons and laser-guided munitions. Air operations in the 21st century could be undertaken both during favourable and unfavourable weather conditions, depending on the availability of reliable technology to do so. Climate change is expected to make a difference towards the conduct of all these operations. The change could have an impact both at the tactical as well as the strategic levels.

Impact at Tactical Level: The impact felt at tactical levels may not be very significant in nature. This is because flying activities are undertaken by giving due cognisance to the prevailing weather conditions and ambient weather inputs; so naturally, any variations in weather would be accounted for. However, the general rise in temperatures expected due to climate change will have some impact on mission planning *per se*. This is because high temperatures reduce lift capability. For any aviation activity, density altitude is a critical measurement that determines if an aircraft has enough lift capability and performance to get off the ground. Too much density altitude limits fuel, weapons and passenger loads. As air density decreases, the altitude increases. The density altitude depends on temperature.

Impact at Strategic Level: At the strategic level, since climatic patterns are likely to show varying changes, the entire mission planning will have to factor in such changes. Particularly, the training activities may suffer if increment in bad weather days (rainy day/thick cloud cover/fog/haze/ strong surface winds) takes place. Also, climatological data of various places is factored in while preparation of existing war plans takes place. Climate change is expected to bring variation in seasonal weather patterns. Naturally, the basic climatic inputs in the preparation of war plans may show change and this needs to be factored in while implementing the existing war plans, or new plans need to be made accordingly.

Recommendations

Military Operations

Glacial Meltdown: Comprehensive and detailed scientific assessments and projections need to be undertaken to estimate the probable extent of glacial melt in the next 10-15 years, especially in areas where the Indian military operates. Adaptation changes in operational and tactical strategies, logistics plans, equipment and infrastructure to the likely meltdown scenario, then need to be worked out in a planning document.

Factoring Long-Term Slow Impacts: The long-term slow impacts of climate change would force gradual adjustments in military operations. Detailed and explicit field studies are required especially with respect to:

- Campaign season.
- Logistics.
- Surface mobility by vehicles, draft animal and on foot.
- Weaponry like change in firing tables and conditions for artillery fire.
- Health impacts due to spread of diseases, both new and old (for both humans and animals).

Naval Operations

- The future security environment and operational conditions being shaped by climate change will affect the Indian Navy the most, being the oceanic global arm.
- Strategic Level: At the strategic level, the Indian Navy will need to revisit its strategic interests and mission types in view of the changing geopolitics, the energy security dynamics, likely future emission constraints, operational conditions, changes in marine biosystems and increased relief, recovery and humanitarian commitments being shaped by climate change. Securing sea lines of communications—an important mission factor—would have newer dimensions. Likely Arctic ocean operations in the future need immediate focus.
- Force Structuring: Increased relief and humanitarian missions, changing operational oceanography, shoreline morphology, extreme weather operations and likely future emission constraints require force structure and decision planning, now envisioning the future scenarios, while balancing present strategic and operational imperatives.
- Sea Level Rise: Sea level rise would affect Indian naval operations at all levels of warfare—strategic, operational and tactical—impinging

on issues of maritime boundaries, EEZ, port operations, shallow water operations—notably for submarines—and naval tactics. For the various scenarios of sea level rise, impact studies need to be undertaken specifically with respect to:

1. Maritime zones and EEZ.
2. Port operations.
3. Protection of bases and coastal installations.

- Ship Design, Equipment and Manning: Ship design with reduced carbon footprint for greater role adaptability for crew and hull for rapid response and deployments, and greater sea-keeping ability to cope with higher sea states and prolonged extreme weather operations. Naval equipment needs to be relevant to climate-induced disaster relief missions especially in construction, sanitation, fuel and water transportation, pharmaceuticals and medical supplies, power generation equipment, etc. Manning and training policies need to factor in carrying humanitarian and civil relief workers more frequently, requiring a shift in operating principles and including them in regular training missions too, in non-crisis periods.

Air Operations

- In general, to maintain a credible military aviation structure, it is essential to cater for changing climatic patterns. This would be only possible if existing structures like the meteorological branches of the Army, Navy and Air Force are trained and equipped to address future challenges. Also, IMD would have to play a key role in respect of data and technology sharing with the armed forces.
- Aerosol concentration in the atmosphere affects the performance of Precision-Guided Munitions (PGMs). Moisture content in the atmosphere, fog particles, dust particles, etc., adversely impact the performance of such weapons. Such minute changes in the aerosol concentration of the atmosphere need to be monitored on a continuous basis. Also, the gaseous concentration in the atmosphere may also change due to climate change. All such types of effects may have a direct or indirect impact on the aviation activities.
- Any military aviation infrastructure constitutes significant ground assets in the form of maintenance services for the aerial platforms,

network of fixed and re-deployable radars, fuel depots, missile silos, etc. Particularly, for the units close to the sea coast, care need to be taken for a likely rise in sea levels.

9 Climate Change: India's Negotiating Position

Arvind Gupta, Sujit Dutta
and Uttam Kumar Sinha

India's negotiating position on climate change has been based on the principles incorporated in the UNFCC. India, like other developing countries, holds the view that the developed countries are responsible to a great degree for the current level of emissions and hence, they should take responsibility for mitigation also. Climate change is taking place due to the accumulated impact of emissions through the process of carbon-based industrialisation for the past 200 years. Even if the emissions are stopped altogether today, global warming will continue for a long time. The narrow focus of the developed countries on current emissions merely distracts from their responsibility for the emissions in the past. Unless this is recognised clearly, the burden of adjustment will fall on developing countries. This is unfair and unjust.

The developing countries have an overriding priority of removing poverty which requires sustained economic growth. For them, emissions are a matter of survival while for the developed countries, emissions are essentially sustaining an unsustainable lifestyle based on unbridled consumption. Thus, the developing countries like India cannot be compelled to take on binding obligations on emission cuts.

In international negotiations, India has argued that the developed countries, following the UNFCC, should take the lead in climate change mitigations. India has pointed out that the emissions of the developed countries, instead of registering a decline, have increased by 2.6 per cent between 2000 and 2005. The emissions of the economies in transition have also begun to show a rising trend. The total emissions of the Annex II countries rose by 11 per cent between 1990 and 2005. India has described these trends as "alarming" and called for their reversal, "so as to release atmospheric space for the development of poorer countries".[1] India is concerned that not only have the GHG emissions of developed countries risen since 2000, their emission in per capita terms has also increased. The emissions of Australia and the US have also risen by 23 per cent and 16 per cent respectively, since 2000. The US emissions should have been 7 per cent below their 1990 level by 2012. While the US emissions are increasing, it is still pressurising the developing countries to reduce their emissions. In India's view, the developed countries should "walk the talk" before they pressurise the developing countries to reduce their emissions.

India believes thc narrow focus by the developed countries on emission reduction has serious repercussions for countries like India. The real fight is for "carbon space" which alone will allow the developing countries to grow. If the developed countries reduce GHG emissions, it will provide impetus for carbon markets and R&D for clean technology. Moreover, there should be no caps on developing countries' emissions, as they are also the potential engines of global growth. In this regard, the continuing growth of China and India should act as an engine of global growth at a time of global economic slowdown. The climate change conditionalities should not cut their growth further.

India has also taken a strong position on the question of adaptation. In India's view, enhancing the implementation of adaptation is high priority for India. India has argued for a new mechanism for adaptation. Fresh sources should be generated for adaptation on the basis of the principles of adequacy, predictability, automaticity and additional funds. India has suggested a simple approach towards adaptation which avoids project-by-project calculations to the extent possible. Adaptation funding should also be predictable in the sense that co-financing levels for different types of interventions in different sectors should be known. Adaptation should also be flexible to accommodate varying country, sector and project-specific conditions. Finally, adaptation funding should be comprehensive to ensure that the needs of a large variety of adaptation interventions can be met.[2]

---

1. Government of India Submission to UNFCCC on MRV-BAP 1(b)(i) available at *http://www.mail.google.com/mail/?ui=781ce O2feO&view=att&th=11 dfad6cf90da62d 12/4/2008*
2. Ibid.

India regards development as the best form of adaptation. It wants infrastructure development to accommodate a climate risk due to climate change, such as flood prevention measures and also strategies to cope with droughts to be regarded as adaptation measures. In India's view, adaptation technologies which help reduce the impact of climate change should be treated at par with mitigation technologies. India has emphasised that mainstreaming climate change concerns into ongoing development programmes in critical sectors should be treated as promoting adaptation.[3]

India has also rejected the "sectoral" approach suggested by Japan. India fears that under the sectoral approach, the developing countries will be forced to use imported technologies and also adopt standards developed by others. This will make developing countries dependent on the developed countries for technology, which will be counter-productive.

India's position has been criticised by several developed countries on the grounds that per capita argument apart, India's rapid growth is adding to the global emissions. India is being described by the developed countries as a major emitter. The US has refused to sign the Kyoto Protocol on the grounds that emerging countries like China, India, Brazil and Russia, who are contributing to the global emissions significantly, have not undertaken any commitments under the Kyoto Protocol. India rejects this argument on the grounds mentioned above. India has sought to persuade the developed countries to provide adequate funds and technology to the developing countries to help adapt to the adverse impact of climate change.

India has responded to the pressure by developed countries by pointing out that it has taken a number of measures which will help reduce emissions. Its National Action Plan on Climate Change (NAPCC) announced in June 2008, includes eight national missions on sustainable development. These include solar, water and wasteland degradation. Likewise, the Indo-US nuclear deal formalised in October 2008 will also help increase the share of non-polluting nuclear energy in India's energy mix. India has implemented an energy efficiency code for new buildings, introduced standards and labelling for energy efficient devices and has a National Environmental Policy, an Energy Conservation Act as well

3. Ibid.

as a new Electricity Act that mandates the procurement of electricity from renewable sources. India has institutional structures to promote sustainable development which will not only ensure growth but also contain emissions.

Kyoto Protocol

The debate on reducing GHG emissions centres on the Kyoto Protocol. A fresh round of negotiations, initiated in Bali in 2007, is expected to finish in Copenhagen in 2009 and will hopefully result in a multilateral agreement which will set targets on emission cuts for the second commitment period, beginning from 2012, within the framework of Kyoto Protocol (See Appendix A). While the negotiation process is gaining momentum, it may be mentioned that there are serious misgivings about the efficacy of the Kyoto Protocol itself. The available data shows that the developed countries are unlikely to meet their emission reduction targets under the Kyoto Protocol by 2012. In fact, the emissions by the developed countries have increased during the 2000-2005 period rather than having been reduced (Emissions as on 2006 are given at Appendix C). The resolve of the EU to reduce their emissions by 20 per cent over the 1990 level by 2020 is also weakening due to the onset of global recession. In fact, the global economic slowdown may result in the reduction of CO_2 emissions because of reduced industrial activity. The economic slowdown will distract the international attention from the long-term impact of climate change. Moreover, the Clean Development Mechanism (CDM) under the Kyoto Protocol has been criticised for its various weaknesses, including the ineffective and unverifiable nature of emissions reduction.

A question must be asked whether the Kyoto Protocol is an effective instrument for emissions reduction in the background of the virtual absence of the political will in reducing CO_2 emissions both in the developed and developing countries. There is a distinct possibility that the current round of international negotiations either may not succeed in reaching an agreement or the agreement reached may not be strong enough to make a meaningful contribution to the reduction of CO_2 emissions. Thus, the following scenarios are a reasonable possibility:

- The developed countries fail to meet their Kyoto Protocol targets by 2012.

- No agreement or only a weak agreement is reached on the commitments for the period after 2012.
- China's position diverges from that of India as, by 2020, having completed the restructuring of its economy, it falls in line with the Western approach.
- The Western countries, in order to pressurise the developing countries, "securitise" the climate change debate, thereby emphasising the security dimension of the issue over its development dimension.
- India is pressurised both by developed as well as developing countries to take binding targets. Even some developing countries, whose per capita emissions may be lower than those of India, particularly the ones which face a serious threat from rising sea levels, may pressurise India to be more flexible so that an agreement on climate change is not thwarted.

India will have to think through its approach to climate change negotiations carefully and make an assessment whether a principled position alone is enough to safeguard its interest, or whether India needs to make certain gestures to push for an acceptable and effective agreement on climate change, which also safeguards India's interests.

India could consider the following:

- India must play an active and constructive role in climate negotiations. It would avoid falling into an NPT-like situation where India, due to its moralistic positions, missed the opportunity to be in the Treaty as a nuclear 'have' country and was subject to technological apartheid for several decades.
- It should continue to follow its principled position at climate change negotiations but show flexibility where possible. It should consider what it can get out of an emerging deal even if it is not a perfect deal.
- It should expedite the implementation of NAPCC in a time-bound and target-oriented manner to be decided by India itself. The targets could be converted into equivalent emission-saved targets and projected to the world appropriately. India should focus, in particular on solar energy, wind energy, biofuels, nuclear energy and energy efficiency.

- It should incorporate environmental concerns into its socio-economic policy not because of external pressure but because it is in the interest of sustainable development. Economic development which results in environmental degradation will not be in India's own interest. Thus, environmental law and regulation in India should have teeth and be enforced in a manner which balances the imperative of economic growth with the compulsion of environmental conservation. Best practices in energy use, building and construction, agriculture, water usage, electricity generation and distribution, automobile sector, waste recycling, etc., should be adopted. While India may not agree with the Japanese sectoral approach, some variant of this approach may be useful for India. We should not close our minds to the adaptable part of the sectoral approach.
- India should be alert to the effects by developed countries to link trade issues with emissions.
- India and China have similar positions on climate change. However, India should watch China's position closely to detect any likely changes in its approach as its own emissions grow and how it responds to Western pressure.
- India should pay top priority to collaborative R&D in green technologies. All necessary fiscal, monetary and institutional incentives should be given to forge a public-private partnership in this area. FDI in renewable sources and clean technologies should be encouraged. India should push for a global R&D fund for clean technologies. UNFCC should be tasked for the management of such a fund.
- India should reject the West's approach of securitising climate change; yet it should be alert to its negative impact on human security and national security. India should use climate change as an opportunity to promote multilateral and regional cooperation, particularly, in the SAARC region. India should use hydro-diplomacy and hydro-solidarity to utilise the hydropower potential of the region to promote regional cooperation for mutual benefit.

- Future negotiations are likely to be centred on a total package of adaptation, mitigation, technology-transfer, etc. The Indian negotiating position should take this into account.
- India's domestic response requires bold measures in the sphere of decentralised energy, carbon costs, water, etc. India should invest in renewable sources of energy like solar and wind energy, for the sake of its own economy. These are essentially mitigation measures which should be projected as such.

10 Conclusion

Arvind Gupta, Sunil Chauhan and P.K. Gautam

This report began with a question: Are there any security implications of climate change for India? The answer is, yes. Climate change is a non-traditional security threat. Its impact on human security is immense. It can lead to loss in biodiversity, breakdown of ecosystems which will have a direct negative impact on human society. Human society is tightly linked with national security.

The Working Group looked at the efforts by the West to 'securitise' the climate change issue. The West's efforts to securitise climate change is driven by its agenda to deflect attention from its own responsibility towards burden-sharing, equity and justice, to pressurise the developing countries to fall in line not only on climate change but on other issues as well. The Working Group agreed with the government's view that climate change should not be treated as a global security issue, and the best way to handle it is as a socio-economic issue within the framework of UNFCC or other UN institutions, except the Security Council.

The Working Group felt that while it was proper to oppose the securitisation of climate change, it would be prudent not to ignore the likely security dimensions of climate change. Climate change will have a serious geopolitical impact. The Arctic meltdown is already resulting in adverse geopolitical consequences, including the possibility of military conflict unless timely steps are taken to prevent it. The possibility of internal or external conflicts due to climate change cannot be ruled out.

The Working Group also deliberated upon the impact of the global financial crisis and the likelihood of global economic recession on climate change. It was felt that if recession sets in, the CO_2 emissions may reduce at least temporarily. How much will the reduction be, will depend on the depth and the duration of the slowdown. But the existing stocks of CO_2 (383 ppm) are enough to continue to push the world towards an uncertain future of climate change. Further, the onset of global recession may weaken the commitment of the developed countries towards emissions reduction; climate change negotiations may run into rough weather. The developed countries' emissions have registered an increase of 9.9 per cent over the 1990 levels.

After surveying the available literature and numerous discussions, the Working Group concluded that climate change will have serious implications for India's security. The impact on India's security, however, will not be direct or linear. Climate change will most likely have a serious impact on human security through water scarcity, reduction in agricultural output, erosion of soil, health and increased frequency of disasters. Climate change will also impose an extra burden on the Indian economy and may slow down economic growth. It could cause intra-state conflict over competing resources.

The changes in the physical environment will impact India's socio-economic habitat. The Working Group went through the available studies on India and South Asia; the chapters in the Report present the main conclusions of the latest studies available on India.

The Working Group felt that climate change and its impact will become a factor in India's relations with its neighbouring countries. Our neighbouring countries will also be affected by climate change as badly as India. Migrations from neighbouring countries to India due to climate change-related factors are possible, which will have undesirable consequences for national security. Sentiments like "sons of the soil" will take root, leading to tensions at the local level. Climate change could also lead to the rise in the number of internally displaced people. Water scarcity issues will assume importance in bilateral relations. Thus, it will be wise from India's point of view, to factor in climate change into its foreign policy, particularly with regard to its neighbours. Our effort should be to use climate change as a reason for cooperation. The foreseeable problem of the likely influx of climate refugees from neighbouring states, especially Bangladesh and Nepal, needs to be discussed jointly with the countries to devise suitable strategies to avoid escalation into intra-state conflict. The Dhaka Declaration on Climate Change is an important step forward. Appropriate initiatives within SAARC also need to be initiated to deliberate on the challenge. It should be followed by concrete steps to implement and strengthen cooperation. Climate change should be included as part of the foreign, security and economic policies.

The Report identified India's key vulnerabilities:

- Future projections of surface warming over India indicate that the annual mean area-averaged surface warming is likely to be between 2.0° and 3.0°C and 3.5° and 5.5°C by the middle and end of the 21st century respectively;
- Trends in sea level rise through observed data of the past 40 years along the Indian coast indicate a possible sea level rise between 1.06 to 1.75 mm per year, consistent with the IPCC rate of 1-2 mm per year.
- Every 10°C rise in temperature would reduce wheat production by 4-5 million tonnes. A loss of 10-40 per cent in crop production has been projected by 2100. The findings also suggest lower yields from dairy cattle and decline in fish breeding, migration and harvests.
- Water scarcity in India will dramatically increase vulnerability and threaten food supplies in India. The availability of water is likely to be altered in significant ways due to climate change. About 67 per cent of the nearly 34,000 sq kilometres of the Himalayan glaciers are receding and in the long run, as the ice diminishes, glacial runoffs and river flows in summer will also go down, leading to severe water shortages in the region
- Up to 50 per cent of the country's flora and fauna could be threatened, with at least a quarter of our biodiversity lost. Nearly 68-77 per cent of Indian forests are likely to undergo a sea change to another forest type by 2085.

A summary of expected effects of extreme events over the Indian subcontinent will be:

Phenomenon and Direction of Trend	**21st Century Likelihood**
Over most land areas, warmer and fewer cold days and nights, warmer and more frequent hot days and nights	Virtually certain (greater than 99% chance)
Warm spells/heat waves. Frequency increases over most areas	Very likely (90–99% chance)
Heavy precipitation events. Frequency increases over most areas	Very likely (90–99% chance)
Area affected by drought increases	Likely (66–90% chance)
Intense tropical cyclone activity	Likely (66–90% chance)
Increased incidence of extreme high sea level (excluding tsunamis)	Likely (66–90% chance)

Source: IPCC Interim Working Group I Report, April 2007; IPCC Synthesis Report, November 2007.

What these imply is that natural disasters are likely to increase both in spread, intensity and frequency. Though it will be very difficult to attribute disasters to climate change alone, climate change will be the key driver and efforts must be made to monitor and disaggregate impacts of natural variations, climate change-related extreme weather events and man-made disasters like drainage congestion. With rapid rise in population, current areas prone to disasters will have a greater impact on larger number of people in future. Along with disasters, tensions and violent conflicts will arise due to both internally displaced persons within India and increased migration flows into India from Bangladesh and Nepal. Climate change would aggravate the severity of natural disasters.

One major recommendation made by the Working Group was, to have an integrated framework to deal with climate change issues and disaster mitigation. There is need for institutional integration, as also for more research in this area.

The Working Group also looked at India's negotiating position. It was felt that the government's position is based on sound principles, equity and justice—common but differentiated responsibility, the polluter pays principle, equitable burden sharing, etc.—outlined in the UNFCC. The government is correct in suggesting a convergence of developing and developed countries' emissions. However, there is a distinct possibility that the developed countries may not live up to their responsibilities and instead treat India as a major emitter and pressurise it to undertake emissions quotas. Not only the developed but also the developing countries might pressurise India to take on binding obligations.

The other possibility is that the Kyoto Protocol may itself come under pressure. The developed countries may not be able to fulfil their emissions commitments for various reasons. Developed countries' emissions during 2000-2005 have increased by 2 per cent instead of decreasing. The Kyoto Protocol may not result in any meaningful reduction in CO_2 emissions. The missions of developed countries are increasing. CDM is being criticised for being ineffective in reducing the GHG emissions. India has been pushing for access to green technologies and for technology transfer. This is the right approach; India can set up a group of like-minded countries. Green technologies can be taken up for discussions while India is negotiating with the various economic corporations.

Under such conditions, the governments may have to take recourse to technological solutions. India must be at the forefront of developing green technologies and become a world leader in at least some technologies. Likewise, the government may wish to incorporate climate change in its socio-economic policies to promote sustainable development.

Climate change cannot be de-linked form the overall issue of energy security and economic growth. A decisive shift to renewable energy sources needs to be made through an appropriate allocation of resources, R&D and fiscal and monetary measures, including incentives such as tax holidays for those industries that follow environment-friendly norms.

The Working Group felt that there would be pressure from the developed countries to apply restrictive rules to trade in environmentally clean technologies. The developed countries may put barriers on import of goods from countries using hazardous technologies. This will have major implications for global trade.

The National Action Plan of Climate Change is a good beginning but its time-bound implementation needs to be ensured. This will require major resource allocation, establishment of institutions and monitoring of the missions.

The government has rejected the Japanese approach for sectoral quotas. Rightly so, as any sectoral obligations could bring international pressure on a developing country like India, to adopt technology standards developed outside. This will increase India's technology dependence on the West. Nevertheless, the Working Group felt that there is merit in the sectoral approach if it is applied within India voluntarily without reference to international obligations. This will improve the implementation of NAPCC.

India needs to improve energy efficiency in the industrial, household and transport sectors, as this would be beneficial for the economy and also contain emissions. Voluntary energy efficiency quotas in various sectors will also strengthen India's position in international negotiations. India will have to make a cost-benefit analysis of no-action *versus* some-action. India needs to spend on voluntary adaptation and mitigation measures in the interest of sustainable development. A range of economic instruments, for instance, carbon taxes, subsidies, etc., can be considered for promoting technology adoption. Insurance cover to include vagaries of slimate such as flooding, crop failures, etc., should be looked into. Similarly, siting of industries, planning for infrastructure and location of health facilities should allow for risks emanating for future climate change impact. As regards mitigation, policy and technological interventions which promote conservation, reforestation, energy efficiency and economic instruments for decarbonisation of energy systems should be considered.

The Working Group Report has looked at the possible adverse impact on the strategy and tactics of Indian armed forces. Climate change may bring about changes in terrain and physical degradation of infrastructure used by the military. The campaigning season may change too. In the absence of any data, the Report could consider this issue in a general manner only. There is urgent need for the armed forces to study the impact of climate change on their operations in a systematic manner.

The consequences of climate change for a developing country like India will be serious under any scenario; human security will be affected. India should be well prepared. Our strategy should not be restricted only to resisting emissions quotas. It should be guided by the desire to make our development sustainable. This cannot be done unless we take a voluntary, mission-focused, target-oriented approach to energy efficiency, renewable energy, resource conservation and a host of other issues. India should use climate change as an opportunity to make socio-economic development more sustainable.

Appendices

Appendix A

Bali Action Plan

The Conference of the Parties,

Resolving to urgently enhance implementation of the Convention in order to achieve its ultimate objective in full accordance with its principles and commitments,

Reaffirming that economic and social development and poverty eradication are global priorities,

Responding to the findings of the Fourth Assessment Report of the Intergovernmental Panel on Climate Change that warming of the climate system is unequivocal, and that delay in reducing emissions significantly constrains opportunities to achieve lower stabilization levels and increases the risk of more severe climate change impacts,

Recognizing that deep cuts in global emissions will be required to achieve the ultimate objective of the Convention and emphasizing the urgency[1] to address climate change as indicated in the Fourth Assessment Report of the Intergovernmental Panel on Climate Change,

1. *Decides* to launch a comprehensive process to enable the full, effective and sustained implementation of the Convention through long-term cooperative action, now, up to and beyond 2012, in order to reach an agreed outcome and adopt a decision at its fifteenth session, by addressing, inter alia:

(a) A shared vision for long-term cooperative action, including a long-term global goal for emission reductions, to achieve the ultimate objective of the Convention, in accordance with the provisions and principles of the Convention, in particular the principle of common but differentiated responsibilities and respective capabilities, and taking into account social and economic conditions and other relevant factors;

(b) Enhanced national/international action on mitigation of climate change, including, inter alia, consideration of:

1. Contribution of Working Group III to the Fourth Assessment Report of the Intergovernmental Panel on Climate Change, Technical Summary, pages 39 and 90, and Chapter 13, page 776.

(i) Measurable, reportable and verifiable nationally appropriate mitigation commitments or actions, including quantified emission limitation and reduction objectives, by all developed country Parties, while ensuring the comparability of efforts among them, taking into account differences in their national circumstances;

(ii) Nationally appropriate mitigation actions by developing country Parties in the context of sustainable development, supported and enabled by technology, financing and capacity-building, in a measurable, reportable and verifiable manner;

(iii) Policy approaches and positive incentives on issues relating to reducing emissions from deforestation and forest degradation in developing countries; and the role of conservation, sustainable management of forests and enhancement of forest carbon stocks in developing countries;

(iv) Cooperative sectoral approaches and sector-specific actions, in order to enhance implementation of Article 4, paragraph 1(c), of the Convention;

(v) Various approaches, including opportunities for using markets, to enhance the cost-effectiveness of, and to promote, mitigation actions, bearing in mind different circumstances of developed and developing countries;

(vi) Economic and social consequences of response measures;

(vii) Ways to strengthen the catalytic role of the Convention in encouraging multilateral bodies, the public and private sectors and civil society, building on synergies among activities and processes, as a means to support mitigation in a coherent and integrated manner;

(c) Enhanced action on adaptation, including, inter alia, consideration of:

(i) International cooperation to support urgent implementation of adaptation actions, including through vulnerability assessments, prioritization of actions, financial needs assessments, capacity-building and response strategies, integration of adaptation actions into sectoral and national planning, specific projects and programmes, means to incentivize the implementation of adaptation actions, and other ways to enable climate-resilient development and reduce vulnerability of all Parties, taking into account the urgent and immediate needs of developing countries that are particularly vulnerable to the adverse effects of climate change, especially the least developed countries and small island developing States, and further taking into account the needs of countries in Africa affected by drought, desertification and floods;

(ii) Risk management and risk reduction strategies, including risk sharing and transfer mechanisms such as insurance;

(iii) Disaster reduction strategies and means to address loss and damage associated with climate change impacts in developing countries that are particularly vulnerable to the adverse effects of climate change;

(iv) Economic diversification to build resilience;

(v) Ways to strengthen the catalytic role of the Convention in encouraging multilateral bodies, the public and private sectors and civil society, building on synergies among activities and processes, as a means to support adaptation in a coherent and integrated manner;

(d) Enhanced action on technology development and transfer to support action on mitigation and adaptation, including, inter alia, consideration of:

(i) Effective mechanisms and enhanced means for the removal of obstacles to, and provision of financial and other incentives for, scaling up of the development and transfer of technology to developing country Parties in order to promote access to affordable environmentally sound technologies;

(ii) Ways to accelerate deployment, diffusion and transfer of affordable environmentally sound technologies;

(iii) Cooperation on research and development of current, new and innovative technology, including win-win solutions;

(iv) The effectiveness of mechanisms and tools for technology cooperation in specific sectors;

(e) Enhanced action on the provision of financial resources and investment to support action on mitigation and adaptation and technology cooperation, including, inter alia, consideration of:

(i) Improved access to adequate, predictable and sustainable financial resources and financial and technical support, and the provision of new and additional resources, including official and concessional funding for developing country Parties;

(ii) Positive incentives for developing country Parties for the enhanced implementation of national mitigation strategies and adaptation action;

(iii) Innovative means of funding to assist developing country Parties that are particularly vulnerable to the adverse impacts of climate change in meeting the cost of adaptation;

(iv) Means to incentivize the implementation of adaptation actions on the basis of sustainable development policies;

(v) Mobilization of public- and private-sector funding and investment, including facilitation of climate-friendly investment choices;

(vi) Financial and technical support for capacity-building in the assessment of the costs of adaptation in developing countries, in particular the most vulnerable ones, to aid in determining their financial needs;

2. *Decides* that the process shall be conducted under a subsidiary body under the Convention, hereby established and known as the Ad Hoc Working Group on Long-term Cooperative Action under the Convention, that shall complete its work in 2009 and present the outcome of its work to the Conference of the Parties for adoption at its fifteenth session;

3. *Agrees* that the process shall begin without delay, that the sessions of the group will be scheduled as often as is feasible and necessary to complete the work of the group, where possible in conjunction with sessions of other bodies established under the Convention, and that its sessions may be complemented by workshops and other activities, as required;

4. *Decides* that the first session of the group shall be held as soon as is feasible and not later than April 2008;

5. *Decides* that the Chair and Vice-Chair of the group, with one being from a Party included in Annex I to the Convention (Annex I Party) and the other being from a Party not included in Annex I to the Convention (non-Annex I Party), shall alternate annually between an Annex I Party and a non-Annex I Party;

6. *Takes note* of the proposed schedule of meetings contained in the annex;

7. *Instructs* the group to develop its work programme at its first session in a coherent and integrated manner;

8. *Invites* Parties to submit to the secretariat, by 22 February 2008, their views regarding the work programme, taking into account the elements referred to in paragraph 1 above, to be compiled by the secretariat for consideration by the group at its first meeting;

9. *Requests* the group to report to the Conference of the Parties at its fourteenth session on progress made;

10. *Agrees* to take stock of the progress made, at its fourteenth session, on the basis of the report by the group;

11. *Agrees* that the process shall be informed by, inter alia, the best available scientific information, experience in implementation of the Convention and its Kyoto Protocol, and processes thereunder, outputs from other relevant intergovernmental processes and insights from the business and research communities and civil society;

12. *Notes* that the organization of work of the group will require a significant amount of additional resources to provide for the participation of delegates from Parties eligible to be funded and to provide conference services and substantive support;

13. *Strongly urges* Parties in a position to do so, in order to facilitate the work of the group, to provide contributions to the Trust Fund for Participation in the UNFCCC Process and the Trust Fund for Supplementary Activities for the purposes referred to in paragraph 12 above and to provide other forms of in kind support such as hosting a session of the group.

Appendix B

Indicative Timetable for Meetings of the Ad Hoc Working Group on Long-term Cooperative Action under the Convention in 2008

SESSION DATES

Session 1: March/April 2008

Session 2: June 2008, in conjunction with the twenty-eighth sessions of the subsidiary bodies

Session 3: August/September 2008

Session 4: December 2008, in conjunction with the fourteenth session of the Conference of the Parties.

IMPORTANT MEETINGS HELD ON NEGOTIATIONS IN 2008 POST-BALI

- Bangkok, March 2008.
- Bonn, Germany, June 2008.
- Accra, Ghana in August 2008.
- Poznan, Poland in December 2008. Fourteenth Meeting of Parties (COP 14). See Press Release post-conference below:

UNITED NATIONS CLIMATE CHANGE CONFERENCE - POZNAŃ, POLAND LAYS FOUNDATION FOR COPENHAGEN DEAL

(Poznań, 12 December 2008) – The United Nations Climate Change Conference - Poznań, Poland ended Friday with a clear commitment from governments to shift into full negotiating mode next year in order to shape an ambitious and effective international response to climate change, to be agreed in Copenhagen at the end of 2009.

Progress was made in the area of technology with the endorsement of the Global Environment Facility's "Poznań Strategic Programme on Technology Transfer". The aim of this programme is to scale up the level of investment by levering private investments that developing countries require both for mitigation and adaptation technologies.

"We will now move to the next level of negotiations, which involves crafting a concrete negotiating text for the agreed outcome," said the President of the conference, Polish Minister of the Environment Maciej Nowicki. Parties agreed that a first draft of the text would be available at a UNFCCC gathering in Bonn in June of 2009.

"In addition to having agreed the work programme for next year, we have cleared the decks of many technical issues," President Nowicki said. "Poznań is the place where the partnership between the developing and developed world to fight climate change has shifted beyond rhetoric and turned into real action," he said.

In that spirit, at Poznań, the finishing touches were put to the Kyoto Protocol's adaptation fund, thereby enabling the fund to receive projects in the course of 2009. Parties agreed that the fund (CDM), fed by a share of proceeds from the Kyoto Protocol's clean development mechanism and voluntary contributions, would have a legal capacity granting developing countries direct access.

However, Parties were unable to reach consensus on scaling up funding for adaptation by agreeing to put a levy on the other two Kyoto mechanisms, Joint Implementation and Emissions Trading.

Together with decisions aimed at streamlining and speeding up the CDM, Parties asked the CDM Executive Board to explore procedures and methodologies that would enhance regional and sub-regional distribution of projects. Parties also asked the Board to assess the implications of including carbon capture and storage projects and extending the eligibility criteria for afforestation and reforestation projects.

A key event at the conference was a ministerial round table on a shared vision for long-term cooperative action on climate change. "Governments have sent a strong political signal that despite the financial and economic crisis, significant funds can be mobilized for both mitigation and adaptation in developing countries with the help of a clever financial architecture and the institutions to deliver the financial support," said Yvo de Boer, Executive Secretary of the United Nations Framework Convention on Climate Change (UNFCCC).

"We now have a much clearer sense of where we need to go in designing an outcome which will spell out the commitments of developed countries, the financial support required and the institutions that will deliver that support as part of the Copenhagen outcome," he added.

Countries meeting in Poznań made progress on a number of issues that are important in the short run—up to 2012—particularly for developing countries, including adaptation, finance, technology and reducing emissions from deforestation and forest degradation.

In addition, the conference discussed in detail the issue of disaster management, risk assessment and insurance, essential to help developing countries cope with the inevitable effects of climate change.

Governments meeting under the Kyoto Protocol agreed that commitments of industrialized countries post-2012 should principally take the form of quantified emission limitation and reduction objectives, in line with the type of emission reduction targets they have assumed for the first commitment period of the protocol.

Including the UN Climate Change Conference in Copenhagen (7-18 December), at least four major UNFCCC gatherings will take place next year, the first two in Bonn, Germany (29 March-08 April and 1-12 June) and the third in August/September.

About the UNFCCC

With 192 Parties, the United Nations Framework Convention on Climate Change (UNFCCC) has near universal membership and is the parent treaty of the 1997 Kyoto Protocol. The Kyoto Protocol has to date 183 member Parties. Under the Protocol, 37 States, consisting of highly industrialized countries and countries undergoing the process of transition to a market economy, have legally binding emission limitation and reduction commitments. The ultimate objective of both treaties is to stabilize greenhouse gas concentrations in the atmosphere at a level that will prevent dangerous human interference with the climate system.

About the CDM

Under the CDM, projects that reduce greenhouse gas emissions in developing countries and contribute to sustainable development can earn certified emission reduction (CER) credits. Countries with a commitment under the Kyoto Protocol buy CERs to cover a portion of their emission reduction commitments under the Treaty. There are currently more than 1240 registered CDM projects in 51 countries, and about another 3000 projects in the project registration pipeline. The CDM is expected to generate more than 2.9 billion CERs by the time the first commitment period of the Kyoto Protocol ends in 2012, each equivalent to one tonne of carbon dioxide.

Appendix C

CO_2 Excluding LULUCF (Land Use, Land Use Change and Forestry) in Gg CO_2 Equivalent of some Annex 1 Countries in Giga Grams (1 Giga Gram=10 Billion Grams)

Sl No.	Country	Base Year (1990)	2000	2006	Change from Base Year to Latest Reported Year (%)
1.	Australia	277,803	350,032	390,436	40.5
2.	Austria	62,085	65,928	77,283	24.5
3.	Belarus	101,947	51,911	59,203	-41.9
4.	Belgium	118,817	123,780	119,107	0.2
5.	Bulgaria	86,269	50,482	55,067	-44.3
6.	Canada	455,999	559,997	560,389	22.9
7.	Croatia	24,069	20,102	23,699	-1.5
8.	Czech Republic	163,865	126,756	127,918	-21.9
9.	Denmark	53,989	54,395	58,925	9.1
10.	Estonia	36,358	15,279	15,972	-56.1
11.	European Community	3,352,984	3,348,821	3,466,315	3.4
12.	Finland	56,710	56,969	68,098	20.1
13.	France	395,597	406,098	408,686	3.3
14.	Germany	1,032,172	883,392	880,253	-14.7
15.	Greece	82,422	103,659	109,666	33.1
16.	Hungary	73,335	59,202	60,389	-29.9
17.	Iceland	2,160	2,758	3,035	40.5
18.	Ireland	32,545	44,847	47,320	45.4
19.	Italy	434,783	464,276	488,039	12.2
20.	Japan	1,144,197	1,256,736	1,273,595	11.3
21.	Latvia	19,157	7,031	8,260	-56.9
22.	Liechtenstein	203.1	227.5	241.6	19.0
23.	Lithuania	36,169	12,085	14,524	-59.8
24.	Luxembourg	12,219	9,040	12,108	-0.9
25.	Monaco	105	113	90	-15.0
26.	Netherlands	159,356	169,649	172,219	8.1
27.	New Zealand	25,382	31,048	36,388	43.4
28.	Norway	34,774	41,577	43,259	24.4
29.	Poland	368,678	320,365	330,524	-29.5
30.	Portugal	43,444	63,412	63,993	47.3
31.	Romania	172,000	95,264	111,011	-42.5
32.	Russian Federation	2,497,016	1,469,002	1,577,689	-36.8
33.	Slovakia	61,838	40,195	39,984	-35.3
34.	Slovenia	14,751	15,223	16,878	3.6
35.	Spain	228,508	307,742	359,627	57.4
36.	Sweden	56,301	53,416	51,515	-8.5
37.	Switzerland	44,558	43,916	45,561	2.3
38.	United Kingdom	590,631	550,255	557,855	-5.5
39.	United States	5,061,388	5,932,184	5,975,096	18.1

Source: UNFCC.

Glossary

Adaptation

Adjustment in natural or human systems in response to actual or expected climatic stimuli or their effects, which moderates harm or exploits beneficial opportunities. Various types of adaptation can be distinguished, including anticipatory, autonomous and planned adaptation.

Adaptation Costs

Costs of planning, preparing for, facilitating and implementing adaptation measures, including transition costs.

Aerosols

A collection of air-borne solid or liquid particles, with a typical size between 0.01 and 10 μm, that reside in the atmosphere for at least several hours. Aerosols may be of either natural or anthropogenic origin. Aerosols may influence climate in two ways: directly through scattering and absorbing radiation, and indirectly through acting as condensation nuclei for cloud formation or modifying the optical properties and lifetime of clouds.

Afforestation

Direct human-induced conversion of land that has not been forested for a period of at least 50 years to forested land through planting, seeding and/or the human-induced promotion of natural seed sources.

Aquifer

A stratum of permeable rock that bears water. An unconfined aquifer is recharged directly by local rainfall, rivers and lakes, and the rate of recharge will be influenced by the permeability of the overlying rocks and soils.

Biodiversity

The total diversity of all organisms and ecosystems at various spatial scales (from genes to entire biomes).

Biofuel

A fuel produced from organic matter or combustible oils produced by plants. Examples of biofuel include alcohol, black liquor from the paper-manufacturing process, wood and soybean oil.

Biomass

The total mass of living organisms in a given area or volume; recently dead plant material is often included as dead biomass. The quantity of biomass is expressed as a dry weight or as the energy, carbon or nitrogen content.

Biosphere

The part of the earth system comprising all ecosystems and living organisms in the atmosphere, on land (terrestrial biosphere) or in the oceans (marine biosphere) including derived dead organic matter, such as litter, soil organic matter and oceanic detritus.

Capacity Building

In the context of climate change, capacity building is developing the technical skills and institutional capabilities in developing countries and economies in transition to enable their participation in all aspects of adaptation to, mitigation of and research on climate change, and in the implementation of the Kyoto Mechanisms, etc.

Carbon Cycle

The term used to describe the flow of carbon (in various forms, e.g., carbon dioxide) through the atmosphere, ocean, terrestrial biosphere and lithosphere.

Carbon Dioxide (CO_2)

A naturally occurring gas fixed by photosynthesis into organic matter. A by-product of fossil fuel combustion and biomass burning, it is also emitted from land use changes and other industrial processes. It is the principal anthropogenic greenhouse gas that affects the earth's radiative balance. It is the reference gas against which other greenhouse gases are measured, thus having a Global Warming Potential of 1.

CDM (Clean Development Mechanism)

The CDM allows greenhouse gas emission reduction projects to take place in countries that have no emission targets under the United Nations Framework Convention on Climate Change (UNFCCC) Kyoto Protocol, yet are signatories.

Cholera

A water borne intestinal infection caused by a bacterium (*Vibrio cholerae*) that results in frequent watery stools, cramping abdominal pain and eventual collapse from dehydration and shock.

Climate

Climate in a narrow sense is usually defined as the 'average weather', or more rigorously, as the statistical description in terms of the mean and variability of relevant quantities over a period of time ranging from months to thousands or millions of years. These quantities are most often surface variables such as temperature, precipitation and wind. Climate in a wider sense is the state, including a statistical description, of the climate system. The classical period of time is 30 years, as defined by the World Meteorological Organization (WMO).

Climate Change

Climate change refers to any change in climate over time, whether due to natural variability or as a result of human activity. This usage differs from that in the United Nations Framework Convention on Climate Change (UNFCCC), which defines 'climate change' as: "a change of climate which is attributed directly or indirectly to human activity that alters the composition of the global atmosphere and which is in addition to natural climate variability observed over comparable time periods". See also climate variability.

Climate Change Commitment

Due to the thermal inertia of the ocean and slow processes in the biosphere, the cryosphere and land surfaces, the climate would continue to change even if the atmospheric composition was held fixed at today's values. Past change in atmospheric composition was held fixed at today's values.

Climate Model

A numerical representation of the climate system based on the physical, chemical and biological properties of its components, their interactions and feedback processes, and accounting for all or some of its known properties. The climate system can be represented by models of varying complexity, i.e., for any one component or combination of components a hierarchy of models can be identified, differing in such aspects as the number of spatial dimensions, the extent to which physical, chemical or biological processes are explicitly represented, or the level at which empirical parameterisations are involved. Coupled Atmosphere/Ocean/General Circulation Models (AOGCMs) provide a comprehensive representation of the climate system.

More complex models include active chemistry and biology. Climate models are applied, as a research tool, to study and simulate the climate but also for operational purposes, including monthly, seasonal and inter-annual climate predictions.

Climate Prediction

A climate prediction or climate forecast is the result of an attempt to produce an estimate of the actual evolution of the climate in the future, e.g., at seasonal, inter-annual or long-term time scales. See also climate projection and climate (change) scenario.

Climate Projection

The calculated response of the climate system to emissions or concentration scenarios of greenhouse gases and aerosols, or radiative forcing scenarios, often based on simulations by climate models. Climate projections are distinguished from climate predictions, in that the former critically depend on the emissions/concentration/radiative forcing scenario used, and therefore, on highly uncertain assumptions of future socio-economic and technological development.

Climate (Change) Scenario

A plausible and often simplified representation of the future climate, based on an internally consistent set of climatological relationships and assumptions of radiative forcing, typically constructed for explicit use as input to climate change impact models. A 'climate change scenario' is the difference between a climate scenario and the current climate.

Climate Sensitivity

The equilibrium temperature rise that would occur for a doubling of CO_2 concentration above pre-industrial levels.

Climate System

The climate system is defined by the dynamics and interactions of five major components: atmosphere, hydrosphere, cryosphere, land surface and biosphere. Climate system dynamics are driven by both internal and external forcing, such as volcanic eruptions, solar variations or human-induced modifications to the planetary radiative balance, for instance via anthropogenic emissions of greenhouse gases and/or land use changes.

Climate Variability

Climate variability refers to variations in the mean state and other statistics (such as standard deviations, statistics of extremes, etc.) of the climate on all temporal and spatial scales beyond that of individual weather events. Variability may be due to natural internal processes within the climate system (internal variability), or to variations in natural or anthropogenic external forcing (external variability). See also climate change.

Communicable Disease

An infectious disease caused by transmission of an infective biological agent (virus, bacterium, protozoan or multicellular macroparasite).

Coral

The term 'coral' has several meanings but is usually the common name for the order *Scleractinia*, all members of which have hard limestone skeletons and which are divided into reef-building and non-reef-building, or cold- and warm-water corals.

Deforestation

Natural or anthropogenic process that converts forest land to non-forest. See afforestation and reforestation.

Dengue Fever

An infectious viral disease spread by mosquitoes, often called breakbone fever because it is characterised by severe pain in the joints and back. Subsequent infections of the virus may lead to dengue hemorrhagic fever (DHF) and dengue shock syndrome (DSS), which may be fatal.

Drought

The phenomenon that exists when precipitation is significantly below normal recorded levels, causing serious hydrological imbalances that often adversely affects land resources and production systems.

Ecosystem

The interactive system formed from

all living organisms and their abiotic (physical and chemical) environment within a given area. Ecosystems cover a hierarchy of spatial scales and can comprise the entire globe, biomes at the continental scale or small, well-circumscribed systems such as a small pond.

Emissions Scenario

A plausible representation of the future development of emissions of substances that are potentially radiatively active (e.g., greenhouse gases, aerosols), based on a coherent and internally consistent set of assumptions about driving forces (such as demographic and socio-economic development, technological change) and their key relationships. In 1992, the IPCC presented a set of emissions scenarios that were used as a basis for the climate projections in the Second Assessment Report. These emissions scenarios are referred to as the IS92 scenarios. In the IPCC Special Report on Emissions Scenarios (SRES), new emissions scenarios—the so-called SRES scenarios—were published.

Endemic

Restricted or peculiar to a locality or region. With regard to human health, endemic can refer to a disease or agent present or usually prevalent in a population or geographical area at all times.

Epidemic

Occurring suddenly in incidence rates clearly in excess of normal expectancy, applied especially to infectious diseases but may also refer to any disease, injury or other health-related event occurring in such outbreaks.

Erosion

The process of removal and transport of soil and rock by weathering, mass wasting and the action of streams, glaciers, waves, winds and underground water.

Food Security

A situation that exists when people have secure access to sufficient amounts of safe and nutritious food for normal growth, development and an active and healthy life. Food insecurity may be caused by the unavailability of food, insufficient purchasing power, inappropriate distribution or inadequate use of food at the household level.

General Circulation Model (GCM)

See climate model.

Glacier

A mass of land ice flowing downhill (by internal deformation and sliding at the base) and constrained by the surrounding topography (e.g., the sides of a valley or surrounding peaks). A glacier is maintained by accumulation of snow at high altitudes, balanced by melting at low altitudes or discharge into the sea.

Globalisation

The growing integration and interdependence of countries worldwide through the increasing volume and variety of cross-border transactions in goods and services, free international capital flows, and the more rapid and widespread diffusion of technology, information and culture.

Greenhouse Effect

The process in which the absorption of infrared radiation by the atmosphere warms the earth. In common parlance, the term 'greenhouse effect' may be used to refer either to the natural greenhouse effect, due to naturally occurring greenhouse gases or to the enhanced (anthropogenic) greenhouse effect, which results from gases emitted as a result of human activities.

Greenhouse Gas

Greenhouse gases are those gaseous constituents of the atmosphere, both natural and anthropogenic, that absorb and emit radiation at specific wavelengths within the spectrum of infrared radiation emitted by the earth's surface, the atmosphere and clouds. This property causes the greenhouse effect. Water vapour (H_2O), carbon dioxide (CO_2), nitrous oxide (N_2O), methane (CH_4) and ozone (O_3) are the primary greenhouse gases in the earth's atmosphere. As well as CO_2, N_2O and CH_4, the Kyoto Protocol deals with the greenhouse gases sulphur hexafluoride (SF_6), hydrofluorocarbons (HFCs) and perfluorocarbons (PFCs).

Gross Domestic Product

Gross domestic product (GDP) is the monetary value of all goods and services produced within a nation.

Gross National Product

Gross national product (GNP) is the monetary value of all goods and services produced in a nation's economy, including income generated abroad by domestic residents but without income generated by foreigners.

Ice Cap

A dome-shaped ice mass covering a highland area that is considerably smaller in extent than an ice sheet.

Ice Sheet

A mass of land ice that is sufficiently deep to cover most of the underlying bedrock topography. An ice sheet flows outwards from a high central plateau with a small average surface slope. The margins slope steeply and the ice is discharged through fast flowing ice streams or outlet glaciers, in some cases into the sea or into ice shelves floating on the sea. There are only two large ice sheets in the modern world —on Greenland and Antarctica, the Antarctic ice sheet being divided into east and west by the Transantarctic Mountains; during glacial periods there were others.

Infectious Disease

Any disease caused by microbial agents that can be transmitted from one person to another or from animals to people. This may occur by direct physical contact, by handling of an object that has picked up infective organisms, through a disease carrier, via contaminated water or by the spread of infected droplets coughed or exhaled into the air.

Kyoto Protocol

The Kyoto Protocol was adopted at the Third Session of the Conference of the Parties (COP) to the UN Framework Convention on Climate Change (UNFCCC) in 1997 in Kyoto, Japan. It contains legally binding commitments, in addition to those included in the UNFCCC. Countries included in Annex B of the Protocol (most member countries of the Organisation for Economic Cooperation and Development (OECD) and those with economies in transition) agreed to reduce their anthropogenic greenhouse gas emissions (CO_2, CH_4, N_2O, HFCs, PFCs and SF_6) by at least 5 per cent below 1990 levels in the commitment period 2008 to 2012. The Kyoto Protocol entered into force on 16 February 2005.

Landslide

A mass of material that has slipped downhill by gravity, often assisted by water when the material is saturated; the rapid movement of a mass of soil, rock or debris down a slope.

Leaching

The removal of soil elements or applied chemicals by water movement through the soil.

Malaria

Endemic or epidemic parasitic disease caused by species of the genus *Plasmodium* (Protozoa) and transmitted by mosquitoes of the genus *Anopheles*; produces bouts of high fever and systemic disorders, affects about 300 million and kills approximately 2 million people worldwide every year.

Millennium Development Goals (MDGs)

A list of ten goals, including eradicating extreme poverty and hunger, improving maternal health and ensuring environmental sustainability, adopted in 2000 by the UN General Assembly, i.e., 191 States, to be reached by 2015. The MDGs commit the international community to an expanded vision of development and have been commonly accepted as a framework for measuring development progress.

Mitigation

An anthropogenic intervention to reduce the anthropogenic forcing of the climate system; it includes strategies to reduce greenhouse gas sources and emissions and enhancing greenhouse gas sinks.

Monsoon

A monsoon is a tropical and sub-tropical seasonal reversal in both the surface winds and associated precipitation.

No Regrets Policy

A policy that would generate net social and/or economic benefits irrespective of whether or not anthropogenic climate change occurs.

Ozone

The triatomic form of oxygen (O_3), a gaseous atmospheric constituent. In the troposphere, it is created both naturally and by photochemical reactions involving gases resulting from human activities (photochemical smog). In high concentrations, tropospheric ozone can be harmful to many living organisms. Tropospheric ozone acts as a greenhouse gas. In the stratosphere, ozone is created by the interaction between solar ultraviolet radiation and molecular oxygen (O_2). Depletion of stratospheric ozone, due to chemical reactions that may be enhanced by climate change, results in an increased ground level flux of ultraviolet (UV) B radiation.

Permafrost

Perennially frozen ground that occurs where the temperature remains below 0°C for several years.

Projection

The potential evolution of a quality or set of quantities, often computed with the aid of a model. Projections are distinguished from predictions in order to emphasise that projections involve assumptions—concerning, for example, future socio-economic and technological developments, that may or may not be realised—and are therefore, subject to substantial uncertainty. See also climate projection and climate prediction.

Reforestation

Planting of forests on lands that have previously contained forests but that have been converted to some other use. For a discussion of the term forest and related terms such as afforestation, reforestation and deforestation, see the IPCC Special Report on Land Use, Land Use Change and Forestry (IPCC, 2000).

Resilience

The ability of a social or ecological system to absorb disturbances while retaining the same basic structure and ways of functioning, the capacity for self-organisation and the capacity to adapt to stress and change.

River Discharge

Water flow within a river channel, for example expressed in m3/s. A synonym for streamflow.

Runoff

That part of precipitation that does not evaporate and is not transpired.

Sea-Level Rise

An increase in the mean level of the ocean. Eustatic sea-level rise is a change in global average sea level brought about by an increase in the volume of the world ocean. Relative sea-level rise occurs where there is a local increase in the level of the ocean relative to the land, which might be due to ocean rise and/or land level subsidence. In areas subject to rapid land level uplift, relative sea level can fall.

Sink

Any process, activity or mechanism that removes a greenhouse gas, an aerosol or a precursor of a greenhouse gas or aerosol from the atmosphere.

SRES

The storylines and associated population, GDP and emissions scenarios associated with the Special Report on Emissions Scenarios (SRES), and the resulting climate change and sea-level rise scenarios. Four families of socio-economic scenario (A1, A2, B1 and B2) represent different world futures in two distinct dimensions: a focus on economic *versus* environmental concerns, and global *versus* regional development patterns.

Stakeholder

A person or an organisation that has a legitimate interest in a project or entity, or would be affected by a particular action or policy.

United Nations Framework Convention on Climate Change (UNFCCC)

The Convention was adopted on 9 May 1992, in New York and signed at the 1992 Earth Summit in Rio de Janeiro by more than 150 countries and the European Community. Its ultimate objective is the 'stabilisation of greenhouse gas concentrations in the atmosphere at a level that would prevent dangerous anthropogenic interference with the climate system'. It contains commitments for all Parties. Under the Convention, Parties included in Annex I aim to return greenhouse gas emissions not controlled by the Montreal Protocol to 1990 levels by the year 2000. The Convention entered in force in March 1994. See also Kyoto Protocol.

Urbanisation

The conversion of land from a natural state or managed natural state (such as agriculture) to cities; a process driven by net rural-to-urban migration through which an increasing percentage of the population in any nation or region come to live in settlements that are defined as 'urban centres'.

Vector

A blood-sucking organism, such as an insect, that transmits a pathogen from one host to another. See also vector-borne diseases.

Vector-Borne Diseases

Disease that are transmitted between hosts by a vector organism (such as a mosquito or tick); e.g., malaria, dengue fever and leishmaniasis.

Vulnerability

Vulnerability is the degree to which a system is susceptible to, and unable to cope with, adverse effects of climate change, including climate variability and extremes. Vulnerability is a function of the character, magnitude and rate of climate change and variation to which a system is exposed, its sensitivity and its adaptive capacity.

Water Stress

A country is water-stressed if the available freshwater supply relative to water withdrawals acts as an important constraint on development. Withdrawals exceeding 20 per cent of renewable water supply have been used as an indicator of water stress. A crop is water-stressed if soil-available water, and thus actual evapotranspiration, is less than potential evapotranspiration demands.

Wetland

A transitional, regularly waterlogged area of poorly drained soils, often between an aquatic and a terrestrial ecosystem, fed from rain, surface water or groundwater. Wetlands are characterised by a prevalence of vegetation adapted for life in saturated soil conditions.